István Gaál

Diophantine Equations and Power Integral Bases

New Computational Methods

Birkhäuser
Boston • Basel • Berlin

István Gaál
University of Debrecen
Institute of Mathematics and Informatics
Debrecen Pf. 12 H-4010
Hungary

Library of Congress Cataloging-in-Publication Data

Gaál, Istvan, 1960-
   Diophantine equations and power integral bases : new computational methods / István Gaál.
      p. cm.
   Includes bibliographical references and index.
   ISBN 0-8176-4271-4 (acid-free paper) — ISBN 3-7643-4271-4 (acid-free paper)
      1. Diophantine equations. 2. Algebraic fields. 3. Bases (Linear topological spaces) I.
   Title.
   QA242.G22 2002
   512'.72–dc21

                                                                    2002025353
                                                                    CIP

AMS Subject Classifications: Primary: 11Y50; Secondary: 11D57, 11D59, 11R33

Printed on acid-free paper
©2002 Birkhäuser Boston

Birkhäuser

ISBN 0-8176-4271-4   SPIN 10851411
ISBN 3-7643-4271-4

Reformatted from author's files by TEXniques, Inc., Cambridge, MA.
Printed and bound by Hamilton Printing Company, Rensselaer, NY.
Printed in the United States of America.

9 8 7 6 5 4 3 2 1

A member of BertelsmannSpringer Science+Business Media GmbH

To Gabi, Zsuzsi and Szilvi

Contents

Preface

One of the classical problems in algebraic number theory, going back, among others, to K. Hensel [He08] and H. Hasse [Ha63], is to decide if an algebraic number field K of degree n has a *power integral basis*, that is, an integral basis of type $\{1, \alpha, \ldots, \alpha^{n-1}\}$. This is equivalent to $\mathbb{Z}_K$ being monogenic, that is of the form $\mathbb{Z}[\alpha]$. The main purpose of this book is to describe *algorithms* for determining generators α of power integral bases. This problem is equivalent to solving the corresponding *index form equations*.

It is important to emphasize that in addition to providing the reader with some efficient algorithms for computing generators of power integral bases, the other goal in this work is to show the development of *constructive (algorithmic) methods for solving diophantine equations*, which has come about as a consequence of a systematic study of index form equations. This has a significant impact on our investigations of power integral bases. Many of these methods can also be applied to solving other types of diophantine equations.

The most straightforward benefit of a power integral basis is to have an easy way of performing arithmetic calculations in $\mathbb{Z}_K$, especially an easy way of multiplication. It also helps to improve minimal polynomials of generating elements of number fields in numerical tables. But it has many other, more or less independent applications; to mention only one quite characteristic application, let us recall the result of B. Kovács and A. Pethő [KP91], who proved that in $\mathbb{Z}_K$ there exists a generalized number system if and only if $\mathbb{Z}_K$ is monogenic; moreover the generators of power integral bases are needed to construct these number systems.

This was one of the reasons it became interesting not only to decide the monogenity of $\mathbb{Z}_K$ but also to determine all possible generators of power integral bases. The algorithm for determining power integral bases in cubic fields was one of the first "real" applications of the methods for solving Thue equations. The simple paper [GS89] on this cubic case is one of the most frequently cited papers of the author, which shows what kind of novelty such computations had at the end of the 1980s.

It has been a great challenge to try to extend the algorithms for computing all generators of power integral bases to higher degree number fields; this was finally successful at least up to degree 5 in general, and for many special higher degree fields up to degree about 9, where we reached the limits of capability of the present methods and the capacity of the present computing machinery. Imagine that for a number field of degree n, the index form equation has $n - 1$ variables and degree $n(n - 1)/2$. This means that the index form equation is mostly (already in quartic fields) a very complicated equation that does not even fit onto one page.

As mentioned above, the algorithms we developed for solving index form equations are clearly applicable and fruitful for also solving other types of decomposable form equations, e.g., the algorithm for solving norm form equations.

On the other hand, in special types of fields, in order to make our computations easier, we investigated the structure of index forms and detected their correspondence with simpler types of equations. This enables us to reduce the index form equation to other important types of diophantine equations such as Thue equations, inhomogeneous Thue equations, relative Thue equations, which are hopefully easier to solve. Since in the past no efficient algorithms were known to solve inhomogeneous Thue equations and relative Thue equations, we developed methods for solving these equations as well. This is another important contribution to the constructive theory of diophantine equations.

More than ten years have passed since we began to study constructive methods for determining power integral bases in algebraic number fields. The material of this book is an outgrowth of several lectures given by the author at various conferences, cf. [Ga91], [Ga96b], [Ga98b], [Ga99], [Ga00a]. Also, some parts have appeared in special university courses held by the author.

The book is organized in the following way. In Chapter 1 we fix our notation, describe the basic concepts, and summarize those important results on power integral bases which (being non-algorithmic) do not fit into subsequent chapters of the book.

In Chapter 2 we collected the main tools for solving our equations. Using *Baker's method* (Section 2.1) we obtain huge upper bounds for the unknown exponents of the corresponding unit equations. These bounds must be *reduced* (Section 2.2) using numerical diophantine approximation techniques based on the LLL basis reduction algorithm. Finally, we *enumerate* the possible small values of the unknown exponents lying under the reduced bounds (Section 2.3). The statements of these Sections are refined versions of formerly used lemmas, formulated in a suitable way for our applications.

In Chapter 3 we survey algorithms for solving simpler types of equations, such as *Thue equations* (Section 3.1), *inhomogenous Thue equations* (Section 3.2), and *relative Thue equations* (Section 3.3). The index form equations in special types of number fields can often be reduced to these types of equations. It is useful to have a uniform discussion of these classical types of equations, containing several very recent results, such as the algorithm for relative Thue equations. In this chapter we also include a (recent) algorithm for solving *norm form equations* (Section 3.4), since they are very close to the above types of equations and use the same tools for their resolution.

Chapter 4 gives a general overview of the structure of *index forms*. Especially, if the field K is a *composite* of its subfields (Section 4.4), we obtain results that make the resolution of the index form equations much easier.

In Chapters 5–8 we describe those algorithms that can be used for *solving index form equations in cubic, quartic, quintic, sextic* number fields, respectively. In addition to several efficient methods that work in special types of fields, we have general algorithms up to degree five. For degree six we can determine the generators of power integral bases only if there are subfields. We also include several *infinite parametric familes of fields* and solve the corresponding index form equations in a parametric form.

In Chapter 9 we consider the problem of *relative power integral bases* in cubic and quartic relative extensions of fields. This is used in Chapter 10 to consider power integral bases in some types of *higher degree number fields*, among others, in octic fields with a quadratic subfield, which are considered as relative quartic extensions of the quadratic subfields.

Finally, in Chapter 11 we provide the results of our computations: *tables of generators of power integral bases* in cubic, quartic, sextic fields. These tables might have many applications.

The reader should have a basic knowledge of algebraic numbers and algebraic number fields. Almost all that is needed can be found in W. Narkiewicz's book [Nark74]. For a more sophisticated algorithmic approach, the reader is referred to the book of M. Pohst and H. Zassenhaus [PZ89].

Debrecen, Hungary   *István Gaál*
March, 2002

Acknowledgements

The author is very grateful to Professor George Anastassiou, consultant editor of Birkhäuser Boston, who encouraged the publication of this book. The author is indepted to Professor Kálmán Győry who introduced him to the theory of diophantine equations.

The influence of Professor Attila Pethő led him in the direction of constructive methods.

The author enjoyed the hospitality of Professor Michael Pohst and the joint work with him as a fellow of the Alexander von Humboldt Foundation as well as several other times when the author could escape from the everyday duties of his home university and devote a couple of days to joint research in Berlin.

This book could not have been written without the support of Ann Kostant, editor of Birkhäuser Boston, who helped the author to organize the material to become a book.

The author also thanks the referees for ideas that improved the presentation and for pointing out several misprints in former versions of the book.

Diophantine Equations and Power Integral Bases

New Computational Methods

1
Introduction

Let K be an algebraic number field of degree n with ring of integers $\mathbb{Z}_K$. The ring $\mathbb{Z}_K$ is called *monogenic* if it is a simple ring extension $\mathbb{Z}[\alpha]$ of $\mathbb{Z}$. In this case $\{1, \alpha, \ldots, \alpha^{n-1}\}$ is an integral basis of K, called a *power integral basis*. Our main task is to develop algorithms for determining all generators α of power integral bases. As we shall see, this algorithmic problem is satisfactorily solved for lower degree number fields (especially for cubic and quartic fields) and there are efficient methods for certain classes of higher degree fields. Our algorithms enable us in many cases to describe all power integral bases also in *infinite parametric families* of certain number fields.

It is a classical problem in algebraic number theory to decide if the ring $\mathbb{Z}_K$ is monogenic, that is if K admits power integral bases. In the 1960s H. Hasse [Ha63] (§25.6., p.438) asked if one could give an arithmetic characterization of those number fields which have power integral bases. The first example of a non-monogenic field was given by R. Dedekind [Ded878].

1.1 Basic concepts

It is useful to recall the following well-known statement (cf. Proposition 2.7 of W. Narkiewicz [Nark74]):

Lemma 1.1.1 *Let $\alpha_1, \ldots, \alpha_n \in \mathbb{Z}_K$ be linearly independent over $\mathbb{Q}$ and set $\mathcal{O} = \mathbb{Z}[\alpha_1, \ldots, \alpha_n]$. Then*

$$D_{K/\mathbb{Q}}(\alpha_1, \ldots, \alpha_n) = J^2 \cdot D_K,$$

where

$$J = (\mathbb{Z}_K^+ : \mathcal{O}^+),$$

$\mathbb{Z}_K^+$ *and* $\mathcal{O}^+$ *are the additive groups of the corresponding modules and* D_K *is the discriminant of the field* K.

Let $\alpha \in \mathbb{Z}_K$ be a primitive element of K, that is $K = \mathbb{Q}(\alpha)$. The *index* of α is defined by the module index

$$I(\alpha) = \left(\mathbb{Z}_K^+ : \mathbb{Z}[\alpha]^+ \right).$$

Obviously, α generates a power integral basis in K if and only if $I(\alpha) = 1$. The *minimal index* of the field K is defined by

$$\mu(K) = \min I(\alpha),$$

the minimum taken for all primitive integers. The *field index* of K is

$$m(K) = \gcd I(\alpha),$$

the greatest common divisor also taken for all primitive integers of K. Monogenic fields have both $\mu(K) = 1$ and $m(K) = 1$, but $m(K) = 1$ is not sufficient for the monogenity.

Let $\{1, \omega_2, \ldots, \omega_n\}$ be an integral basis of K. Let

$$L(\underline{X}) = X_1 + \omega_2 X_2 + \cdots + \omega_n X_n$$

with conjugates $L^{(i)}(\underline{X}) = X_1 + \omega_2^{(i)} X_2 + \cdots + \omega_n^{(i)} X_n$, $(i = 1, \ldots, n)$. The form $L(\underline{X})$ was called the "Fundamentalform" and

$$D_{K/\mathbb{Q}}(L(\underline{X})) = \prod_{1 \leq i < j \leq n} \left(L^{(i)}(\underline{X}) - L^{(j)}(\underline{X}) \right)^2$$

the "Fundamental-diskriminante" by L. Kronecker [Kr882] and K. Hensel. [He08]

Lemma 1.1.2 *We have*

$$D_{K/\mathbb{Q}}(L(\underline{X})) = (I(X_2, \ldots, X_n))^2 D_K, \tag{1.1}$$

where D_K *is the discriminant of the field* K *and* $I(X_2, \ldots, X_n)$ *is a homogeneous form in* $n - 1$ *variables of degree* $n(n - 1)/2$ *with integer coefficients.*

This form $I(X_2, \ldots, X_n)$ is called the *index form* corresponding to the integral basis $\{1, \omega_2, \ldots, \omega_n\}$.

Proof. Define $\ell(\underline{X}) = \omega_2 X_2 + \cdots + \omega_n X_n$ with conjugates $\ell^{(i)}(\underline{X}) = \omega_2^{(i)} X_2 + \cdots + \omega_n^{(i)} X_n$. Obviously, there are homogeneous polynomials $F_{ji} \in \mathbb{Z}[X_2, \ldots, X_n]$ $(1 \leq j \leq n)$ of degree $i - 1$ such that

$$\left(\ell(\underline{X}) \right)^{i-1} = F_{1i}(\underline{X}) + F_{2i}(\underline{X}) \omega_2 + \cdots + F_{ni}(\underline{X}) \omega_n$$

holds for $1 \leq i \leq n$. Further,

$$
\begin{aligned}
D_{K/\mathbb{Q}}(L(\underline{X})) &= \prod_{1 \leq i < j \leq n} \left(L^{(i)}(\underline{X}) - L^{(j)}(\underline{X}) \right)^2 \\
&= \prod_{1 \leq i < j \leq n} \left(\ell^{(i)}(\underline{X}) - \ell^{(j)}(\underline{X}) \right)^2 \\
&= \begin{vmatrix} 1 & \ell^{(1)}(\underline{X}) & \cdots & \left(\ell^{(1)}(\underline{X})\right)^{n-1} \\ \vdots & \vdots & & \vdots \\ 1 & \ell^{(n)}(\underline{X}) & \cdots & \left(\ell^{(n)}(\underline{X})\right)^{n-1} \end{vmatrix}^2 \\
&= \begin{vmatrix} 1 & \omega_2^{(1)} & \cdots & \omega_n^{(1)} \\ \vdots & \vdots & & \vdots \\ 1 & \omega_2^{(n)} & \cdots & \omega_n^{(n)} \end{vmatrix}^2 \cdot \begin{vmatrix} F_{11}(\underline{X}) & \cdots & F_{1n}(\underline{X}) \\ \vdots & & \vdots \\ F_{n1}(\underline{X}) & \cdots & F_{nn}(\underline{X}) \end{vmatrix}^2 \\
&= D_K \cdot (I(X_2, \ldots, X_n))^2,
\end{aligned}
$$

where the degree of the homogeneous polynomial

$$
I(X_2, \ldots, X_n) = \begin{vmatrix} F_{11}(\underline{X}) & \cdots & F_{1n}(\underline{X}) \\ \vdots & & \vdots \\ F_{n1}(\underline{X}) & \cdots & F_{nn}(\underline{X}) \end{vmatrix}
$$

is $n(n-1)/2$, which can be calculated from the degrees of the polynomials $F_{ji}(\underline{X})$. $\square$

A very important property of the index form is formulated in the following lemma:

Lemma 1.1.3 *For any primitive integer* $\alpha = x_1 + \omega_2 x_2 + \cdots + \omega_n x_n \in \mathbb{Z}_K$,

$$
I(\alpha) = |I(x_2, \ldots, x_n)|
$$

holds.

Proof. By Lemma 1.1.1 we have

$$
D_{K/\mathbb{Q}}(\alpha) = D_{K/\mathbb{Q}}(1, \alpha, \ldots, \alpha^{n-1}) = (I(\alpha))^2 \cdot D_K.
$$

On the other hand, Lemma 1.1.2 implies

$$
D_{K/\mathbb{Q}}(\alpha) = D_{K/\mathbb{Q}}(L(x_1, \ldots, x_n)) = (I(x_2, \ldots, x_n))^2 \cdot D_K.
$$

$\square$

Obviously, α generates a power integral basis of K if and only if $(x_2, \ldots, x_n) \in \mathbb{Z}^{n-1}$ satisfies the *index form equation*

$$
I(x_2, \ldots, x_n) = \pm 1 \quad \text{in} \quad x_2, \ldots, x_n \in \mathbb{Z}. \tag{1.2}
$$

Hence the problem of determining all power integral bases in K is equivalent to solving equation (1.2).

The index form is independent of the variable X_1. If $\beta = a \pm \alpha$ for any $a \in \mathbb{Z}$, then the indices of α and β are equal. Such algebraic integers are called *equivalent*.

In Chapter 9 we shall investigate power integral bases in relative extensions of number fields. If K has a subfield M, then the *relative index* of $\alpha \in \mathbb{Z}_K$ (such that $K = M(\alpha)$) with respect to the extension K/M is

$$I_{K/M}(\alpha) = (\mathbb{Z}_K^+ : \mathbb{Z}_M^+[\alpha]).$$

As we shall see, if α generates a power integral basis in K, then it must have relative index 1 with respect to M.

1.2   Related results

Using Baker's method, K. Győry [Gy76] gave the first explicit upper bounds for the absolute values of the solutions of equation (1.2), see Theorem 2.1.2. These bounds imply that up to equivalence there are only finitely many generators of power integral bases. Unfortunately these bounds are much too large for practical applications. The finiteness of the number of solutions of equation (1.2) follows also from the result of B.J. Birch and J.R. Merriman [BM71].

Combining the effective bounds obtained by Baker's method with reduction methods (see Section 2.2) and enumeration algorithms (see Section 2.3) we shall be able to determine the solutions of the index form equation (1.2) for lower degree fields and for some classes of higher degree fields.

In this section we give a brief overview of former (mainly non–algorithmic) results on the monogenity of rings of integers of number fields.

In case of cubic fields B.N. Delone [Del30] and T. Nagell [Nag30] proved the finiteness of the number of solutions of equation (1.2) using Thue's result [Thu09] (cf. Theorem 3.1.2). In cyclic cubic number fields the problem of monogenity was considered by M.N. Gras [Gr73], [Gr75], and G. Arnichard [Ar74] who gave necessary and sufficient conditions for the monogenity. D.S. Dummy and H. Kisilevsky [DK77] showed that infinitely many totally real cyclic cubic fields have power integral bases, and on the other hand the minimum index of such fields can be arbitrarily large. The monogenity of relative cyclic cubic extensions of quadratic fields was studied by J.D. Thérond [The93], [The95].

For quartic fields T. Nagell [Nag67], [Nag68] showed the finiteness of the number of solutions of equation (1.2). In case of cyclic quartic fields M.N. Gras [Gr79] gave necessary and sufficient conditions of the monogenity. She showed that imaginary cyclic quartic fields can have power integral bases only in two trivial cases. T. Nakahara [Nak87] proved that the minimum index of cyclic quartic fields can be arbitrarily large. Concerning quartic fields of type $K = \mathbb{Q}(\sqrt{m}, \sqrt{n})$, (those with Galois group V_4, the Klein four group) T. Nakahara [Nak83] showed that infinitely many of such fields admit power integral bases, but their minimal index can be arbitrarily large. M.N. Gras and F. Tanoe [GT95] gave necessary and sufficient conditions for the monogenity of such fields.

The non-monogenity of certain types of cyclic quartic and cyclic sextic fields was shown by T. Nakahara [Nak93].

M.N. Gras [Gr86] proved the following statement on cyclic fields of prime degree:

Theorem 1.2.1 *If K is a cyclic field of prime degree ≥ 5, then $\mathbb{Z}_K$ has no power integral bases except when K is the maximal real subfield of a cyclotomic field.*

Generalizing the method used to prove the above theorem, M.N. Gras [Gr83] showed:

Theorem 1.2.2 *Let $n \geq 5$ be relatively prime to 2 and 3. There are only finitely many fields of degree n with Abelian Galois group having power integral basis.*

There are interesting results on power integral bases in rings of integers of cyclotomic fields. Let p be an odd prime, ζ a primitive p-th root of unity. A. Bremner [Br88] conjectured that up to equivalence only ζ and $1/(1 + \zeta)$ generate a power integral basis in the cyclotomic field $\mathbb{Q}(\zeta)$. The statement was proved by L. Robertson [Ro98] for $p \leq 23$, $p \neq 17$, for $p = 17$ this was shown by K. Wildanger [Wi00]. If ζ is a primitive 2^m-th root of unity, then L. Robertson [Ro01] proved that up to equivalence all generators of power integral bases in the cyclotomic field $\mathbb{Q}(\zeta)$ are given by ζ^i, where $i \in \mathbb{Z}$ is odd. Observe that this interesting result characterizes power integral bases in fields of arbitrarily large degrees.

We finally recall some results on the field index $m(K)$. Denote by $v_p(m(K))$ the greatest exponent of a prime p dividing $m(K)$. R. Dedekind [Ded878] showed that $v_p(m(K)) > 1$ if and only if there exist an f such that the number of prime ideal divisors of p in $\mathbb{Z}_K$ of degree f is greater than the number of irreducible polynomials of degree f over the field of p elements. O. Ore [Or28] conjectured that although by Dedekind's result $v_p(m(K)) = 0$ or > 0 depends only on the decomposition type of p in K, this is not true for the value of $v_p(m(K))$. H.T. Engstrom [En30] showed that for number fields of degree $n \leq 7$ the value of $v_p(m(K))$ depends only on the decomposition type of p in K and computed $v_p(m(K))$ for all these fields. He also proved Ore's conjecture by finding an example for degree $n = 8$. J. Sliwa [Sl82] proved that if p is unramified in K, then $v_p(m(K))$ depends only on the decomposition type of p in K. E. Nart [Nart85] described which arithmetic invariants of K determine the value of $v_p(m(K))$.

There are two problems in W. Narkiewicz's book [Nark74] related to this topic. Problem No. 6 asks us to find a good necessary and sufficient condition for a field to have index 1. Problem No. 22 asks for an explicit formula for $v_p(m(K))$.

2
Auxiliary Results, Tools

In this Chapter we summarize the basic tools we use all through the book to solve our diophantine equations. The equations are usually reduced to so-called *unit equations* in two variables of type

$$\alpha u + \beta v = 1$$

(cf. equation (2.5)) with given algebraic α, β, where u, v are unknown units in a number field. These units are written as a power product of the generators of the unit group and the unknown exponents are to be determined. Baker's method (Section 2.1) is used to give an initial upper bound for the unknowns, which is of magnitude 10^{18} for the simplest Thue equations but 10^{100} is also not unusual for more complicated equations. We apply numerical diophantine approximation techniques based on the LLL basis reduction algorithm (Section 2.2) to reduce these bounds. The reduced bound is usually between 100 and 1000. These reduced bounds are quite modest, however if there are more than 4–5 of them, it is already impossible to test directly all possible exponents with absolute values under the reduced bound. Hence we have to apply certain enumeration methods (Section 2.3) to overcome this difficulty.

In fact we can observe that the majority of computations are devoted to proving that there are no "large" solutions. According to our experience the solutions are usually "small" and can be found sometimes quite easily.

2.1   Baker's method, effective finiteness theorems

Beginning in the mid 1960s A. Baker [Ba90] developed a powerful method deriving lower estimates for certain linear forms in the logarithms of algebraic numbers. This method was successfully applied to several types of diophantine equations [Ba90]. For our purposes we recall here an improvement of these estimates obtained by A. Baker and G. Wüstholz [BW93] that fits better to our applications and is also much sharper than the original estimates.

Let α be an algebraic number of degree $\deg \alpha$. Let $a_0 \in \mathbb{Z}$ be the leading coefficient of the minimal polynomial of α and denote by $\alpha^{(i)}$, $(i = 1, \ldots, \deg \alpha)$ the conjugates of α. By the usual notation

$$h(\alpha) = \frac{1}{\deg \alpha} \left(\log a_0 + \sum_{i=1}^{\deg \alpha} \log \max(1, |\alpha^{(i)}|) \right)$$

is the *absolute logarithmic height* of α.

Let $\alpha_1, \ldots, \alpha_n$ be algebraic numbers, different from 0 and 1, and let $\log \alpha_1, \ldots, \log \alpha_n$ be the principal value of their logarithms. Let $K = \mathbb{Q}(\alpha_1, \ldots, \alpha_n)$ be of degree d. For $1 \le i \le n$ set

$$H_i = \max \left\{ h(\alpha_i), \frac{1}{\deg \alpha_i}, \frac{|\log \alpha_i|}{\deg \alpha_i} \right\}.$$

Let $a_1, \ldots, a_n \in \mathbb{Z}$, not all 0, and consider the linear form

$$\Lambda = a_1 \log \alpha_1 + \cdots + a_n \log \alpha_n.$$

Theorem 2.1.1 *If $\Lambda \ne 0$, then*

$$\log |\Lambda| > -C(n, d) H_1 \cdots H_n \log A, \tag{2.1}$$

where $A = \max(|a_1|, \ldots, |a_n|, e)$ and

$$C(n, d) = 18(n + 1)! n^{n+1} (32d)^{n+2} \log(2nd).$$

By using inequalities of the above type, K. Győry [Gy76] deduced the first effective upper bounds for the solutions of equation (1.2). These bounds have had many improvements, here we recall the recent result of K. Győry [Gy98].

Let $\{1, \omega_2, \ldots, \omega_n\}$ be an integral basis of the number field K. Assume $\overline{|\omega_i|} \le H$ for $i = 2, \ldots, n$, where $\overline{|\gamma|}$ denotes the *size* of an algebraic number γ, that is the maximum absolute value of its conjugates.

Let $K^{(i)}$ $(i = 1, \ldots, n)$ be the conjugate fields of K. Denote by n_2 the maximum of the degrees of $K^{(i)} K^{(j)}$ $(1 \le i < j \le n)$ and by n_3 the maximum of the degrees of $K^{(i)} K^{(j)} K^{(l)}$ $(1 \le i < j < l \le n)$. Note that $n_2 \le n(n - 1)$ and $n_3 \le n(n - 1)(n - 2)$.

Theorem 2.1.2 *Every solution of the index form equation (1.2) satisfies*

$$\max_{2 \le i \le n} |x_i| \; < \; \exp \left\{ 3^{n_2+29} n_3^{9n_2+15} |D_K|^{n_2/n} \right.$$
$$\left. \times (\log |D_K|)^{2n_2-1} \left(|D_K|^{n_2/n} + \log H \right) \right\}.$$

The type of estimates like (2.1) implies that the constants here are double exponential functions of some parameters occurring in the formula. This results in a huge bound even in the simplest special cases. These bounds imply that up to equivalence there are only finitely many generators of power integral bases in the field K. Unfortunately these bounds are much too large for practical applications.

We shall use reduction methods (see Section 2.2) to get a reasonable upper bound for the variables and then apply enumeration algorithms (see Section 2.3) to determine the solutions of the index form equation (1.2) for lower degree fields and some classes of higher degree fields.

2.2   Reduction

We first deal with the simplest case, originally used by Baker and Davenport, then we pass to the general case.

2.2.1   Davenport lemma

To give an impression of the nature of these numerical algorithms we first give a short discussion of Davenport's lemma [BD69] since all further methods originate from it. Solving, e.g., a cubic Thue equation with real roots, standard arguments (cf. Section 3.1) lead to an inequality of type

$$|a_1\theta + a_2 - \gamma| < C \cdot K^{-A} \tag{2.2}$$

with real numbers θ, γ, positive constants C, K of moderate size, and unknown exponents $a_1, a_2 \in \mathbb{Z}$ with $A = \max(|a_1|, |a_2|)$. Baker type inequalities yield an upper bound A_0 for A, which is usually of magnitude 10^{18} in such cases.

Lemma 2.2.1 *Let M, B $(B > 6)$ be given positive integers. If there exists an integer $q \in \mathbb{Z}$, with*

$$(i) \qquad 1 \le q \le MB,$$
$$(ii) \qquad \|q\theta\| < \frac{2}{MB},$$
$$(iii) \qquad \|q\gamma\| > \frac{3}{B},$$

then (2.2) has no solutions a_1, a_2 with

$$\frac{\log(MB^2C)}{\log K} \le A \le M, \tag{2.3}$$

where $\|.\|$ denotes the distance from the nearest integer.

This is a slight improvement (cf. I. Gaál [Ga88]) of the lemma of A. Baker and H. Davenport [BD69], usually cited as Davenport's lemma, introducing the constant C that makes the result more efficient.

Proof. Assuming $A \le M$ we show that $A < \log(MB^2C)/\log K$.
Let p be the nearest integer to $q\vartheta$, then we have

$$|a_1 (q\vartheta - p)| < M \frac{2}{MB} = \frac{2}{B}.$$

Let $(-p_0)$ be the nearest integer to $(a_1 (\vartheta q - p) - q\gamma)$. Then

$$\begin{aligned}
\frac{1}{B} = \frac{3}{B} - \frac{2}{B} &< \|q\gamma\| - |a_1 (q\vartheta - p)| \\
&\le |q\gamma - p_0| - |a_1 (\vartheta q - p)| \\
&\le |a_1 (\vartheta q - p) - q\gamma + p_0| \\
&= \|a_1 (\vartheta q - p) - q\gamma\| \\
&\le |a_1 (\vartheta q - p) + a_1 p + a_2 q - q\gamma| \\
&= |a_1 q\vartheta + a_2 q - \gamma q| \le |q| \, CK^{-A} \le MBCK^{-A},
\end{aligned}$$

whence the assertion follows. $\square$

Set $M = A_0$, and $B = 1000$ (which is usually a suitable choice). Observe that in the logarithmic expression of (2.3) M is dominating, B, C, K usually have moderate values. Hence, if we calculate an appropriate $q \in \mathbb{Z}$, then in view of the lemma we can reduce A_0 almost to its logarithm. The q satisfying (i) and (ii) is obtained by the continued fraction expansion of ϑ, namely one of the denominators of the convergents to ϑ is suitable. This q will usually also satisfy (iii), because γ is independent from ϑ (otherwise replace B by $10B$ for example). The reduction step can be repeated as long as the reduced bound is smaller than the previous one. Finally (after about 4–5 reduction steps), the bound is usually smaller than 100, and it is easy to enumerate and test all possible values of a_1, a_2 below this bound.

2.2.2 The general case

Solving more complicated equations, we usually get estimates of type

$$|a_1\xi_1 + \cdots + a_n\xi_n| < c_1 \exp(-c_2 A - c_3), \tag{2.4}$$

where $\xi_1, \ldots, \xi_n$ are real numbers, linearly independent over $\mathbb{Q}$ and c_1, c_2, c_3 are given positive constants (of moderate size). Our purpose is to reduce the bound A_0 obtained previously for $A = \max(|a_1|, \ldots, |a_n|)$ by using Baker's method.

Generalization of the Davenport lemma to the case of several variables was given by A. Pethő and R. Schulenberg [PS87] and B.M.M. de Weger [We89]. The following statement that fits better to our applications was used by I. Gaál and M. Pohst [GP96], [GP01], I. Gaál and K. Győry [GGy99], and I. Gaál [Ga01], [Ga00c].

Let H be a large constant (to be specified later) and consider the lattice $\mathcal{L}$ spanned by the columns of the $n+2$ by n matrix

$$
\begin{pmatrix}
1 & 0 & \cdots & 0 \\
0 & 1 & \cdots & 0 \\
\vdots & \vdots & \vdots & \vdots \\
0 & 0 & \cdots & 1 \\
H\mathfrak{R}(\xi_1) & H\mathfrak{R}(\xi_2) & \cdots & H\mathfrak{R}(\xi_n) \\
H\mathfrak{I}(\xi_1) & H\mathfrak{I}(\xi_2) & \cdots & H\mathfrak{I}(\xi_n)
\end{pmatrix}.
$$

Denote by b_1 the first vector of an LLL-reduced basis of this lattice (cf. A.K. Lenstra, H.W. Lenstra Jr. and L. Lovász [LLL82], and M. Pohst [Po93]).

Lemma 2.2.2 *If* $A \le A_0$ *and* $|b_1| \ge \sqrt{(n+1)2^{n-1}} \cdot A_0$, *then*

$$
A \le \frac{\log H + \log c_1 - c_3 - \log A_0}{c_2}.
$$

Proof. Denote by l_0 the shortest vector in the lattice. Using the inequalities of [Po93] we have $|b_1|^2 \le 2^{n-1}|l_0|^2$. Then by the assumptions, using also (2.4),

$$
2^{1-n}\left((n+1)\cdot 2^{n-1}A_0^2\right)
$$
$$
\le 2^{1-n}|b_1|^2 \le |l_0|^2 \le n \cdot A_0^2 + H^2 c_1^2(\exp(-c_2 A - c_3))^2 ,
$$

whence

$$
A_0 \le H c_1 \exp(-c_2 A - c_3) ,
$$

which implies the assertion.   □

If the terms in (2.4) are linearly dependent over $\mathbb{Q}$, then we can reduce the number of variables. If $\xi_1, \ldots, \xi_n$ are real, we can omit the imaginary parts in the last component of the generating vectors of the lattice $\mathcal{L}$.

An appropriate value of H corresponding to A_0 is of magnitude A_0^n. The algorithm is fast even for 6–7 variables and still feasible for 10–11 variables since the LLL reduction algorithm is fast.

The reduction is very efficient, in the first step the former bound is reduced almost to its logarithm. After about 4–5 steps, the procedure stabilizes, i.e., the new bound is not any smaller than the previous one. Then we stop the procedure. The final reduced bound is usually between 100 and 1000.

2.3   Enumeration methods

In our applications we shall meet equations of type

$$\beta^{(I)} + \beta^{(I')} = 1, \tag{2.5}$$

where I, I' are multi-indices (denoting usually conjugates), from a certain set $\mathcal{I}$, having the property

$$\text{if } I \in \mathcal{I}, \qquad \text{then there are } I', I'' \in \mathcal{I} \text{ with}$$
$$\beta^{(I')} = 1 - \beta^{(I)}, \quad \beta^{(I'')} = \frac{1}{\beta^{(I)}}. \tag{2.6}$$

These numbers are of the form

$$\beta^{(I)} = \tau^{(I)} \left(v_1^{(I)} \right)^{a_1} \dots \left(v_n^{(I)} \right)^{a_n}, \tag{2.7}$$

where $\tau, v_1, \dots, v_n$ are given algebraic numbers, $v_1, \dots, v_n$ are multiplicatively independent. We are going to determine the variables $a_1, \dots, a_n \in \mathbb{Z}$. In fact (2.5) is a type of *unit equation*. In our applications these equations have a special structure, especially the unknown exponents are the same in both terms on the left-hand side.

Using Baker's method and the reduction algorithm we can derive an upper bound A_R for $A = \max(|a_1|, \dots, |a_n|)$. This reduced bound is of magnitude 100 in most cases. If the number n of variables is at most 4, we can try to enumerate and test directly the possible values of exponents. If $n = 4, 5$ the application of sieve methods may help. However, for $n > 5$ these methods are not feasible any more.

K. Wildanger [Wi97], [Wi00] has developed a very efficient enumeration algorithm that solves the above problem and can be successfully used up to 10–11 variables. He showed that instead of an n-dimensional cube containing all possible values of the variables with absolute values under the reduced bound, using the information contained in equation (2.5), we only have to enumerate the integer points in some thin ellipsoids, which can be efficiently done by using the method of U. Fincke and M. Pohst [FP85].

He described the method for general unit equations, here we give a version of it that fits better to our applications. This version was used e.g., by I. Gaál and M. Pohst [GP01], I. Gaál and K. Győry [GGy99], I. Gaál [Ga01], [Ga00c].

Let $I^* = \{I_1, \dots, I_t\}$ be a subset of $\mathcal{I}$ having minimal number of elements with the following properties:

1. If $I \in I^*$, then

   - either $I' \in I^*$ with $\beta^{(I')} = 1 - \beta^{(I)}$,
   - or $I'' \in I^*$ with $\beta^{(I'')} = 1/(1 - \beta^{(I)})$.

2. If $I \in I^*$, then

- either $I' \in I^*$ with $\beta^{(I')} = 1 - 1/\beta^{(I)} = (\beta^{(I)} - 1)/\beta^{(I)}$,
- or $I'' \in I^*$ with $\beta^{(I'')} = \beta^{(I)}/(\beta^{(I)} - 1)$.

3. The vectors

$$
\underline{e}_h = \begin{pmatrix} \log \left| v_h^{(I_1)} \right| \\ \vdots \\ \log \left| v_h^{(I_t)} \right| \end{pmatrix} \quad \text{for} \quad h = 1, \dots, n
$$

are linearly independent over $\mathbb{Q}$.

Condition (2.6) on $\mathcal{I}$ ensures that there are suitable indices $I', I'' \in \mathcal{I}$ satisfying Conditions in 1 and 2.

Since $v_1, \dots, v_n$ are multiplicatively independent, taking sufficiently many indices I, Condition 3 can be satisfied. Note that taking a minimal set of multi-indices satisfying the above conditions reduces the amount of necessary computation considerably.

Set

$$
\underline{g} = \begin{pmatrix} \log \left| \tau^{(I_1)} \right| \\ \vdots \\ \log \left| \tau^{(I_t)} \right| \end{pmatrix}, \quad \underline{b} = \begin{pmatrix} \log \left| \beta^{(I_1)} \right| \\ \vdots \\ \log \left| \beta^{(I_t)} \right| \end{pmatrix}.
$$

By our notation we have

$$
\underline{b} = \underline{g} + a_1 \underline{e}_1 + \cdots + a_n \underline{e}_n. \tag{2.8}
$$

Denote by A_R the reduced bound obtained by Baker's method and reduction for $A = \max(|a_1|, \cdots, |a_n|)$. Let

$$
\log S_0 = \max_{I \in I^*} \left(\left| \log \left| \tau^{(I)} \right| \right| + A_R \left| \log \left| v_1^{(I)} \right| \right| + \ldots + A_R \left| \log \left| v_n^{(I)} \right| \right| \right). \tag{2.9}
$$

Then in view of (2.7) for any $I \in I^*$ we have

$$
\frac{1}{S_0} \leq \left| \beta^{(I)} \right| \leq S_0. \tag{2.10}
$$

In the applications S_0 is very large, ca. of magnitude 10^{1000}.

For our applications concerning equation (2.5) I. Gaál and M. Pohst [GP01] constructed a suitable version of the method of Wildanger [Wi97], [Wi00]. It describes how to replace S_0 by a smaller constant.

Lemma 2.3.1 *Let* $2 < s < S$ *be given constants and assume that*

$$
\frac{1}{S} \leq \left| \beta^{(I)} \right| \leq S \quad \text{for all} \quad I \in I^*.
$$

Then either

$$
\frac{1}{s} \leq \left| \beta^{(I)} \right| \leq s \quad \text{for all} \quad I \in I^*, \tag{2.11}
$$

or there is an $I \in I^*$ with

$$\left| \beta^{(I)} - 1 \right| \leq \frac{1}{s - 1}.$$

Proof. Assume that the tuple $I \in I^*$ violates (2.11). Then either

$$\frac{1}{S} \leq |\beta^{(I)}| \leq \frac{1}{s},$$

which implies by (2.5) that for I' in Condition 1. we have

$$\left| \beta^{(I')} - 1 \right| \leq \frac{1}{s}. \tag{2.12}$$

On the other hand, if

$$s \leq \left| \beta^{(I)} \right| \leq S,$$

then for I' in Condition 2 we have

$$\left| \beta^{(I')} - 1 \right| = \left| \frac{1}{\beta^{(I)}} \right| \leq \frac{1}{s}. \tag{2.13}$$

If in either case $I' \notin I^*$ but $I'' \in I^*$, then using $\beta^{(I'')} = 1/\beta^{(I')}$ by (2.12) and (2.13) we have

$$\left| \beta^{(I'')} - 1 \right| \leq \frac{1}{s - 1}.$$

$\square$

Summarizing, the constant S can be replaced by the smaller constant s if for each t_0 $(1 \leq t_0 \leq t)$ we enumerate directly the set H_{t_0} of those exponents $a_1, \dots, a_n$ for which

$$\frac{1}{S} \leq \left| \beta^{(I)} \right| \leq S \quad \text{for all} \quad I \in I^*$$

and $\tag{2.14}$

$$\left| \beta^{(I_{t_0})} - 1 \right| \leq \frac{1}{s - 1}.$$

We consider the enumeration of the above set H_{t_0} in detail, this being the critical step of the algorithm. Assume that $2 < s < S$ and set

$$\lambda_p = \begin{cases} \dfrac{1}{\log S} & \text{for} \quad p \neq t_0, \ 1 \leq p \leq t, \\[3mm] \dfrac{1}{\log \frac{s-1}{s-2}} & \text{for} \quad p = t_0, \\[3mm] \dfrac{1}{\arccos \frac{s(s-2)}{(s-1)^2}} & \text{for} \quad p = t + 1. \end{cases}$$

Set

$$\varphi_{t_0}(\underline{b}) = \begin{pmatrix} \lambda_1 \log \left| \beta^{(I_1)} \right| \\ \vdots \\ \lambda_t \log \left| \beta^{(I_t)} \right| \\ \lambda_{t+1} \arg \left(\beta^{(I_{t_0})} \right) \end{pmatrix}, \quad \varphi_{t_0}(\underline{g}) = \begin{pmatrix} \lambda_1 \log \left| \tau^{(I_1)} \right| \\ \vdots \\ \lambda_t \log \left| \tau^{(I_t)} \right| \\ \lambda_{t+1} \arg \left(\tau^{(I_{t_0})} \right) \end{pmatrix}$$

and

$$\varphi_{t_0}(\underline{e}_h) = \begin{pmatrix} \lambda_1 \log \left| v_h^{(I_1)} \right| \\ \vdots \\ \lambda_t \log \left| v_h^{(I_t)} \right| \\ \lambda_{t+1} \arg \left(v^{(I_{t_0})} \right) \end{pmatrix} \qquad \text{for } h = 1, \dots, n,$$

where $-\pi \le \arg z \le \pi$ for $z \in \mathbb{C}$, and

$$\underline{e}_0 = \begin{pmatrix} 0 \\ \vdots \\ 0 \\ \lambda_{t+1}\pi \end{pmatrix} \in \mathbb{R}^{t+1}.$$

Since $\underline{e}_1, \dots, \underline{e}_n$ are linearly independent over $\mathbb{Q}$, so are the images $\varphi_{t_0}(\underline{e}_1), \cdots, \varphi_{t_0}(\underline{e}_n)$ as well, and (2.8) implies that there exists an integer a_0 such that

$$\varphi_{t_0}(\underline{b}) = \varphi_{t_0}(\underline{g}) + a_1 \varphi_{t_0}(\underline{e}_1) + \ldots + a_n \varphi_{t_0}(\underline{e}_n) + a_0 \underline{e}_0.$$

We deduce from (2.14) (cf. the elementary inequalities of [Wi97] Lemma 1.13) that

$$\left| \log \left| \beta^{(I_p)} \right| \right| \le \begin{cases} \log S & \text{if} \quad p \ne t_0, \\ \log \frac{s-1}{s-2} & \text{if} \quad p = t_0, \end{cases}$$

and

$$\left| \arg \beta^{(I_{t_0})} \right| \le \arccos \frac{s(s-2)}{(s-1)^2}.$$

Consequently, for the norm of the vector $\varphi_{t_0}(\underline{b})$ we have

$$\|\varphi_{t_0}(\underline{g}) + a_1 \varphi_{t_0}(\underline{e}_1) + \cdots + a_n \varphi_{t_0}(\underline{e}_n) + a_0 \underline{e}_0\|_2^2 = \|\varphi_{t_0}(\underline{b})\|_2^2$$
$$= \sum_{p=1}^{t} \lambda_p^2 \log^2 \left| \beta^{(I_p)} \right| + \lambda_{t+1}^2 \arg^2 \left(\beta^{(I_{t_0})} \right) \le t+1. \qquad (2.15)$$

Hence we have shown that for any $(a_1, \dots, a_n) \in H_{t_0}$ the inequality (2.15) holds. This inequality defines an *ellipsoid*. The lattice points contained in this ellipsoid can be enumerated by using the algorithm of U. Fincke and M. Pohst [FP85]. The enumeration is usually very fast, but it is essential that the "improved" version (cf. [FP85]) of the algorithm should be used, involving LLL reduction.

If all components involved in our formulas are real, then we can omit the $(t+1)$-th component of the vectors, the variable a_0 and the vector $\underline{e}_0$. In such cases we have t instead of $t+1$ on the left-hand side of (2.15).

The vector $\underline{g}$ can be dependent on the vectors $\underline{e}_1, \ldots, \underline{e}_n$ over $\mathbb{R}$ or independent from them. If it is dependent, then (in the real case),

$$\underline{g} = r_1 \cdot \underline{e}_1 + \cdots + r_n \cdot \underline{e}_n$$

for certain real numbers $r_1, \ldots, r_n$. That is, in view of (2.15) we have to enumerate the solutions of the form $y_h = a_h + r_h$ $(1 \le h \le n)$ of the ellipsoid

$$\|y_1 \varphi_{t_0}(\underline{e}_1) + \cdots + y_n \varphi_{t_0}(\underline{e}_n)\|_2^2 \le t$$

and from the values of y_h we determine the a_h. This makes a bit more complicated some processes involved in the Fincke–Pohst algorithm.

Applying the above procedure we choose appropriate constants $S_0 > S_1 > \cdots > S_k$. In each step we take $S = S_i, s = S_{i+1}$ and enumerate the lattice points in the corresponding ellipsoids. The initial constant is given by the reduced bound (2.10); the last constant S_k should be made as small as possible, so that the exponents with

$$\frac{1}{S_k} \le \left| \beta^{(I)} \right| \le S_k \quad \text{for all} \quad I \in I^* \tag{2.16}$$

can be enumerated easily. Observe that the set (2.16) is also contained in an ellipsoid, namely, by (2.8) we have in $\mathbb{R}^t$

$$\|\underline{g} + a_1 \underline{e}_1 + \cdots + a_n \underline{e}_n\|_2^2 = \|\underline{b}\|_2^2 \le t \cdot (\log S_k)^2. \tag{2.17}$$

For an optimal choice of the constants S_i see K. Wildanger [Wi97]. According to our experience in the first step S_1 can be much smaller than S_0, e.g., $S_1 = 10^{20}$. Then it is economical to have $S_{i+1} = \sqrt{S_i}$ until S_i decreases to about 10^3. Then we choose $S_{i+1} = S_i/2$.

2.4   Software, hardware

One of our main purposes is to perform computations in number fields using the algorithms described here. For all these algorithms the basic data of the number fields involved (such as integral bases, fundamental units) are considered as input data. These can either be taken from tables, or can be calculated by using algebraic number theory packages. We mostly used KASH [DF97] but one can also use PARI [Co93] . The methods can be applied to solving similar types of decomposable form equations, as well.

Most of the algorithms were implemented in Maple [CG88]. Implementing them in a more efficient system, e.g., in KASH in would result in better running times, but Maple is very convenient in developing the algorithms, trying different ideas which often helps to improve the procedures.

The algorithms were executed mostly on an IBM PC compatible machine with a PII 350MHz processor in a Linux system.

3

Auxiliary Equations

Index form equations can often be reduced to simpler types of diophantine equations. This is the case e.g., when the index form factorizes. As we shall see in the following chapters, various types of Thue equations play an essential role in the resolution of index form equations [Ga96b]. We summarize the methods for the resolution of these equations in this chapter. We shall consider Thue equations (Section 3.1), inhomogeneous Thue equations (Section 3.2) and relative Thue equations (Section 3.3). The algorithms for solving them are based uniformly on the tools of the preceeding chapter. In this chapter we also include an algorithm for solving certain types of norm form equations (Section 3.4), the type of the equation and the ideas for solving it being very close to what we use for the various types of Thue equations.

3.1 Thue equations

The resolution of index form equations in cubic and quartic number fields is based on solving Thue equations (cf. Sections 5.1, 6.1). Since these equations are treated in detail by N. Tzanakis and B.M.M. de Weger [TW89], [TW92], B.M.M. de Weger [We89], A. Pethő [Pe99] and N. Smart [Sm98], we only give here a short overview of the methods for solving these equations. Note that the program package KASH [DF97] contains a reliable routine for solving Thue equations.

3.1.1 Elementary estimates

Let $F \in \mathbb{Z}[X, Y]$ be an irreducible homogeneous form of degree $n \geq 3$ and let m be a non-zero integer. Consider the Thue equation

$$F(x, y) = m \quad \text{in} \quad x, y \in \mathbb{Z}. \tag{3.1}$$

We can assume that the leading coefficient of F is 1, otherwise if the leading coefficient is a, we multiply the equation by a^{n-1}, replace (x, y) by (ax, ay) and get the required situation.

Let $\alpha^{(1)}, \ldots, \alpha^{(n)}$ be the roots of $F(x, 1) = 0$, (these are conjugates of an algebraic integer $\alpha = \alpha^{(1)}$) and let (x, y) be an arbitrary but fixed solution of (3.1). Then equation (3.1) can be written in the form

$$\prod_{j=1}^{n}(x - \alpha^{(j)}y) = m. \tag{3.2}$$

Let $\beta^{(j)} = x - \alpha^{(j)}y$ $(1 \leq j \leq n)$ and denote by i the index with $|\beta^{(i)}| = \min_j |\beta^{(j)}|$. Obviously $|\beta^{(i)}| \leq \sqrt[n]{|m|}$, hence

$$
\begin{aligned}
|\beta^{(j)}| \geq |\beta^{(j)} - \beta^{(i)}| - |\beta^{(i)}| &\geq |\alpha^{(j)} - \alpha^{(i)}||y| - \sqrt[n]{|m|} \\
&\geq \frac{1}{2}|\alpha^{(j)} - \alpha^{(i)}||y|,
\end{aligned}
\tag{3.3}
$$

for all $j \neq i$ if $|y| > 2\sqrt[n]{|m|}/|\alpha^{(j)} - \alpha^{(i)}|$ (small values of y can be directly checked). Using (3.2), inequality (3.3) implies

$$\left|\alpha^{(i)} - \frac{x}{y}\right| < \frac{c_1}{|y|^n} \tag{3.4}$$

with

$$c_1 = \frac{\sqrt[n]{|m|}}{\prod_{j \neq i} |\alpha^{(j)} - \alpha^{(i)}|}.$$

3.1.2 Thue's theorem

Equation (3.1) is called Thue's equation in honour of A. Thue who gave [Thu09] an ineffective improvement of Liouville's theorem in 1909.

Theorem 3.1.1 *If α is an algebraic integer of degree n, then for every $\varepsilon > 0$ there exists a positive constant $c = c(\alpha, \varepsilon)$, such that for every rational number x/y,*

$$\left|\alpha - \frac{x}{y}\right| > \frac{c}{|y|^{n/2+1+\varepsilon}}$$

holds.

Thue's proof does not make it possible to calculate the constant c. The above inequality together with (3.4) implies the existence of an upper bound for $|y|$, which can not be calculated explicitly. This yields Thue's famous theorem [Thu09]:

Theorem 3.1.2 *Equation (3.1) has only finitely many solutions.*

The method of Thue does not enable one to find the solutions.

3.1.3   Fast algorithm for finding "small" solutions

Consider now the solutions of the *Thue inequality*

$$|F(x, y)| \le m \quad \text{in} \quad x, y \in \mathbb{Z}, \ (x, y) = 1, \ |y| < C, \tag{3.5}$$

where C is a given large constant, say 10^{500}. For our further purposes we mention here an efficient algorithm of A. Pethő [Pe87] producing the solutions of (3.5). In practice the method gives all solutions but does not give a proof of the non-existence of solutions with $|y| \ge C$.

Observe that the inequalities (3.3) and (3.4) remain valid also in the present situation. If $|y|$ is not very small, then (3.4) implies

$$\left| \alpha^{(i)} - \frac{x}{y} \right| < \frac{c_1}{|y|^n} < \frac{1}{2|y|^2}.$$

If $\alpha^{(i)}$ is real (otherwise the solutions are easy to determine by considering the imaginary part of $(x - \alpha^{(i)} y)$), Legendre's theorem implies that x/y is a convergent p_k/q_k in the continued fraction algorithm of $\alpha^{(i)}$. Using another well-known inequality, if a_k is the corresponding partial quotient, then

$$\frac{1}{(a_{k+1} + 2)q_k^2} < \left| \alpha^{(i)} - \frac{p_k}{q_k} \right|,$$

that is

$$\frac{1}{(A + 2)q_k^2} \le \frac{1}{(a_{k+1} + 2)q_k^2} < \left| \alpha^{(i)} - \frac{p_k}{q_k} \right| < \frac{c_1}{|q_k|^n},$$

with $A = \max_{1 \le k \le k_0} a_k$, where k_0 is the first index for which $q_{k_0} > C$. Hence the above inequality implies

$$|y| = q_k < (c_1(A + 2))^{1/(n-2)}. \tag{3.6}$$

Summarizing: $|y|$ is either small, or x/y is a convergent in the continued fraction expansion to a real conjugate of α with (3.6). Thus in order to solve (3.5) we have to test small values of y, calculate the maximum of the indicated partial quotients to the real conjugates of α and test the convergents satisfying (3.6). Inequality (3.6) implies also a kind of reduction of the bound C: having the bound in (3.6) we can repeat the calculation using the bound in (3.6) instead of C.

3.1.4 Effective methods

Return now to equation (3.1). The first effective bounds for the solutions of Thue equations were given by A. Baker (see [Ba90]), the best known bounds are due to Y. Bugeaud and K. Győry [BGy96b].

Let $\eta_1, \ldots, \eta_r$ be a set of fundamental units of $K = \mathbb{Q}(\alpha)$ and determine a full set of non-associated elements μ of norm $\pm m$ in $\mathbb{Z}_K$. These data can be computed by KASH [DF97]. The following must be performed for all μ in this finite set (usually for small norms there are only some candidates). We have

$$\beta = \xi \cdot \mu \cdot \eta_1^{a_1} \cdots \eta_r^{a_r} \tag{3.7}$$

with a root of unity ξ and $a_1, \ldots, a_r \in \mathbb{Z}$ with $A = \max(|a_1|, \ldots, |a_r|)$. Taking conjugates, absolute values and logarithms in (3.7), we get a system of linear equations in $a_1, \ldots, a_r$,

$$a_1 \log \left| \eta_1^{(j)} \right| + \cdots + a_r \log \left| \eta_r^{(j)} \right| = \log \left| \frac{\beta^{(j)}}{\mu^{(j)}} \right|, \tag{3.8}$$

where the conjugate j runs through all non-equivalent embeddings of K into $\mathbb{C}$ excluding the i-th conjugate. By using Cramer's rule we obtain

$$A \le c_2 \max \left| \log \left| \frac{\beta^{(j)}}{\mu^{(j)}} \right| \right| \le c_3 \log |y|, \tag{3.9}$$

where c_2 is the row norm of the inverse matrix of $(\log |\eta_k^{(j)}|)$, that is the maximum sum of the absolute values of the elements in a row, and $c_3 = 2c_2$ if $|y|$ is large enough. Finally, take j, k such that i, j, k are distinct, then Siegel's identity

$$(\alpha^{(i)} - \alpha^{(j)})\beta^{(k)} + (\alpha^{(j)} - \alpha^{(k)})\beta^{(i)} + (\alpha^{(k)} - \alpha^{(i)})\beta^{(j)} = 0$$

gives

$$\frac{(\alpha^{(i)} - \alpha^{(j)})\beta^{(k)}}{(\alpha^{(i)} - \alpha^{(k)})\beta^{(j)}} - 1 = \frac{(\alpha^{(k)} - \alpha^{(j)})\beta^{(i)}}{(\alpha^{(i)} - \alpha^{(k)})\beta^{(j)}}. \tag{3.10}$$

Using the representation (3.7) and the estimates (3.3), (3.4), (3.9) the above equation implies

$$
\begin{aligned}
\Lambda = {} & \left| \log \left| \frac{(\alpha^{(i)} - \alpha^{(j)})\mu^{(k)}}{(\alpha^{(i)} - \alpha^{(k)})\mu^{(j)}} \right| + a_1 \log \left| \frac{\eta^{(k)}}{\eta^{(j)}} \right| + \cdots \right. \\
& \left. + a_r \log \left| \frac{\eta^{(k)}}{\eta^{(j)}} \right| \right| \\
\le {} & 2 \left| \left| \frac{(\alpha^{(i)} - \alpha^{(j)})\beta^{(k)}}{(\alpha^{(i)} - \alpha^{(k)})\beta^{(j)}} \right| - 1 \right| \le 2 \left| \frac{(\alpha^{(i)} - \alpha^{(j)})\beta^{(k)}}{(\alpha^{(i)} - \alpha^{(k)})\beta^{(j)}} - 1 \right| \\
= {} & 2 \left| \frac{(\alpha^{(k)} - \alpha^{(j)})\beta^{(i)}}{(\alpha^{(i)} - \alpha^{(k)})\beta^{(j)}} \right| \le c_4 |y|^{-n} \\
= {} & c_4 \exp(-n \log |y|) \le c_4 \exp(-c_5 A). \tag{3.11}
\end{aligned}
$$

Here we used the elementary inequality $|\log x| < 2|x - 1|$, holding for $|x - 1| < 0.795$, which is ensured if A is not very small,

$$c_4 = 4c_1 \left| \frac{\alpha^{(j)} - \alpha^{(k)}}{(\alpha^{(i)} - \alpha^{(k)})(\alpha^{(i)} - \alpha^{(j)})} \right|, \quad c_5 = \frac{n}{c_3}.$$

If the terms in the above linear form in the logarithm of certain algebraic numbers are linearly independent over $\mathbb{Q}$, then using Theorem 2.1.1 of A. Baker and G. Wüstholz we can give a lower bound of the form

$$\exp(-C \log A) < \Lambda$$

for Λ with a huge positive constant C. Comparing it with (3.11) we get an upper bound A_0 for A. This bound is very large, about 10^{18} even for the simplest cubic equations, 10^{35} for quartic equations etc.

By (3.11), namely from

$$\left| \log \left| \frac{(\alpha^{(i)} - \alpha^{(j)})\mu^{(k)}}{(\alpha^{(i)} - \alpha^{(k)})\mu^{(j)}} \right| + a_1 \log \left| \frac{\eta_1^{(k)}}{\eta_1^{(j)}} \right| + \cdots + a_r \log \left| \frac{\eta_r^{(k)}}{\eta_r^{(j)}} \right| \right| \le c_4 \exp(-c_5 A)$$

we can derive an inequality of type (2.4). Using Lemma 2.2.2 we can reduce the large upper bound for A. The reduced bound A_R for A is under 100 for small units ranks (e.g., 2,3) and is under 1000 for larger unit ranks.

Having the reduced bound for A we can either derive an upper bound for $|y|$ by (3.7) and use the method of Section 3.1.3 to find the "small" solutions, or by (3.10) we get the unit equation

$$\frac{\left(\alpha^{(i)} - \alpha^{(j)}\right) \xi^{(k)} \mu^{(k)}}{\left(\alpha^{(i)} - \alpha^{(k)}\right) \xi^{(j)} \mu^{(j)}} \left(\frac{\eta_1^{(k)}}{\eta_1^{(j)}} \right)^{a_1} \cdots \left(\frac{\eta_r^{(k)}}{\eta_r^{(j)}} \right)^{a_r}$$

$$+ \frac{\left(\alpha^{(j)} - \alpha^{(k)}\right) \xi^{(i)} \mu^{(i)}}{\left(\alpha^{(i)} - \alpha^{(k)}\right) \xi^{(j)} \mu^{(j)}} \left(\frac{\eta_1^{(i)}}{\eta_1^{(j)}} \right)^{a_1} \cdots \left(\frac{\eta_r^{(i)}}{\eta_r^{(j)}} \right)^{a_r} = 1,$$

to which the enumeration methods of Section 2.3 can be applied. In the later case (3.7) enables one to calculate x, y from the exponents $a_1, \ldots, a_r$.

3.1.5   The method of Bilu and Hanrot

The most efficient method for solving Thue equations is due to Y. Bilu and G. Hanrot [BH96]. If (u_{kj}) is the inverse matrix of $\left(\log \left| \eta_k^{(j)} \right| \right)$ used above, then (3.8) implies

$$a_k = \sum_{j=1}^{r} u_{kj} \log \left| \frac{\beta^{(j)}}{\mu^{(j)}} \right| = \sum_{j=1}^{r} u_{kj} \log \left| \frac{x - \alpha^{(j)} y}{\mu^{(j)}} \right|$$

$$= \left(\sum_{j=1}^{r} u_{kj} \right) \log |y| + \sum_{j=1}^{r} u_{kj} \log \left| \frac{\alpha^{(j)} - \dfrac{x}{y}}{\alpha^{(i)} - \alpha^{(j)}} \right| + \sum_{j=1}^{r} u_{kj} \log \left| \frac{\alpha^{(i)} - \alpha^{(j)}}{\mu^{(j)}} \right|$$

(where in the summation j runs through all non-equivalent embeddings of K into $\mathbb{C}$ excluding the i-th one) which in view of

$$\log \left| \frac{\alpha^{(j)} - \frac{x}{y}}{\alpha^{(i)} - \alpha^{(j)}} \right| = \log \left| \frac{\alpha^{(i)} - \frac{x}{y}}{\alpha^{(i)} - \alpha^{(j)}} - 1 \right| < \frac{2}{|\alpha^{(i)} - \alpha^{(j)}|} \left| \alpha^{(i)} - \frac{x}{y} \right|,$$

(3.4) and (3.9) imply

$$|\delta_k \log |y| - a_k + v_k| < c_6 \exp(-c_5 A)$$

with

$$\delta_k = \sum_{j=1}^{r} |u_{kj}|, \quad v_k = \left| \sum_{j=1}^{r} u_{kj} \log \left| \frac{\alpha^{(i)} - \alpha^{(j)}}{\mu^{(j)}} \right| \right|, \quad c_6 = 2c_1 \cdot \sum_{j=1}^{r} \left| \frac{u_{kj}}{\alpha^{(i)} - \alpha^{(j)}} \right|.$$

Taking any two of these inequalities (say for k, l) and eliminating the term with $\log |y|$ we obtain an inequality of the form

$$|a_k \vartheta + a_l - \delta| < c_7 \exp(-c_5 A) \le c_7 \exp(-c_5 \max(|a_k|, |a_l|))$$

with

$$\vartheta = -\frac{\delta_l}{\delta_k}, \quad \delta = \frac{\delta_k v_l - \delta_l v_k}{\delta_k}, \quad c_7 = \frac{c_6(|\delta_k| + |\delta_l|)}{|\delta_k|}.$$

To this inequality the Davenport lemma (see Lemma 2.2.1) can be used. Thus both the reduction and enumeration processes become very easy. This enables one to solve Thue equations of very high degree as well.

3.2 Inhomogeneous Thue equations

Let α be an algebraic integer of degree $n \ge 3$, $K = \mathbb{Q}(\alpha)$, let $0 \ne m \in \mathbb{Z}$. In some applications for index form equations in sextic and octic fields (cf. Sections 8.1.1, 8.1.2, 10.1.3) we shall need to solve equations of type

$$N_{K/\mathbb{Q}}(x + \alpha y + \lambda) = m \quad \text{in} \quad x, y \in \mathbb{Z}, \lambda \in \mathbb{Z}_K \qquad (3.12)$$

where we assume that $\overline{|\lambda|} < X^{1-\zeta}$, where $X = \max(|x|, |y|)$ and $0 < \zeta < 1$ is a given constant. (As usual, $\overline{|\lambda|}$ denotes the size of λ that is the maximum absolute value of its conjugates.) V.G. Sprindžuk [Sp74] considered equations of this type. This equation might also be referred to as a *inhomogeneous Thue equation*. Using Baker's method he gave effective upper bounds for the solutions of equation (3.12). The variables x, y are called *dominating variables*, while λ is called *non-dominating variable*. In [Ga88] we gave a numerical method of solving (3.12), which we briefly explain below. Equations of similar structure will be applied e.g., for solving index form equations corresponding to some sextic and octic fields. (In fact in those cases λ will be bounded by an absolute constant, making the resolution much easier.)

3.2.1 Elementary estimates

Let (x, y, λ) be a fixed solution of equation (3.12). Set $\beta = x + \alpha y + \lambda$. Denote by $\beta^{(i)}$ the conjugate of β with $|\beta^{(i)}| = \min |\beta^{(j)}|$. Obviously

$$|\beta^{(i)}| \le \sqrt[n]{|m|}. \tag{3.13}$$

For $j \ne i$ we have

$$|\beta^{(j)}| \ge |\beta^{(j)} - \beta^{(i)}| - |\beta^{(i)}| \ge |\alpha^{(j)} - \alpha^{(i)}||y| - 2X^{1-\varsigma} - \sqrt[n]{|m|}.$$

On the other hand

$$\alpha^{(i)}\beta^{(j)} - \alpha^{(j)}\beta^{(i)} = (\alpha^{(i)} - \alpha^{(j)})x + \alpha^{(i)}\lambda^{(j)} - \alpha^{(j)}\lambda^{(i)},$$

whence

$$|\beta^{(j)}| \ge \frac{1}{|\alpha^{(i)}|} \left(|\alpha^{(i)} - \alpha^{(j)}||x| - (|\alpha^{(i)}| + |\alpha^{(j)}|)X^{1-\varsigma} - |\alpha^{(j)}|\sqrt[n]{|m|} \right).$$

The later two inequalities imply

$$|\beta^{(j)}| \ge c_1 X - c_2 X^{1-\varsigma} - c_3 > c_4 X \tag{3.14}$$

if X is large enough (the opposite case is easy to consider), where

$$c_1 = \min \left(|\alpha^{(j)} - \alpha^{(i)}|, \frac{|\alpha^{(i)} - \alpha^{(j)}|}{|\alpha^{(i)}|} \right),$$

$$c_2 = \max \left(2, \frac{|\alpha^{(i)}| + |\alpha^{(j)}|}{|\alpha^{(i)}|} \right),$$

$$c_3 = \sqrt[n]{|m|} \, \max \left(1, \frac{|\alpha^{(j)}|}{|\alpha^{(i)}|} \right),$$

$$c_4 = \frac{c_1}{2},$$

the constants c_1, c_2, c_3, c_4 depending of course on i, j. From inequality (3.14) we infer

$$|\beta^{(i)}| < \frac{|m|}{\prod_{j \ne i} |\beta^{(j)}|} < c_5 X^{1-n} \tag{3.15}$$

with

$$c_5 = \frac{|m|}{\prod_{j \ne i} c_4(i, j)}.$$

Let $\eta_1, \ldots, \eta_r$ be the fundamental units of $K = \mathbb{Q}(\alpha)$ and consider a full set of non-associated elements μ of norm $\pm m$ in $\mathbb{Z}_K$. These data can be computed by KASH [DF97]. The following must be performed for all μ in this finite set. We have

$$\beta = \xi \, \mu \, \eta_1^{a_1} \cdots \eta_r^{a_r} \tag{3.16}$$

with a root of unity ξ and $a_1, \ldots, a_r \in \mathbb{Z}$ with $A = \max(|a_1|, \ldots, |a_r|)$. Similarly as for Thue equations this equation implies

$$A \le c_6 \max \left| \frac{\beta^{(j)}}{\mu^{(j)}} \right| \le c_7 \log X, \tag{3.17}$$

where c_6 is the row norm of the inverse matrix of $(\log |\eta_k^{(j)}|)$ used also in the preceding sections and $c_7 = 2c_6$ if X is large enough.

3.2.2   Baker's method

Let i, j, k be distinct indices. Siegel's identity takes the form

$$(\alpha^{(i)} - \alpha^{(j)})(\beta^{(k)} - \lambda^{(k)}) + (\alpha^{(j)} - \alpha^{(k)})(\beta^{(i)} - \lambda^{(i)})$$
$$+ (\alpha^{(k)} - \alpha^{(i)})(\beta^{(j)} - \lambda^{(j)}) = 0,$$

whence

$$\frac{(\alpha^{(i)} - \alpha^{(j)})\beta^{(k)}}{(\alpha^{(i)} - \alpha^{(k)})\beta^{(j)}} - 1 = \frac{(\alpha^{(j)} - \alpha^{(k)})\beta^{(i)} + \delta}{(\alpha^{(i)} - \alpha^{(k)})\beta^{(j)}} \tag{3.18}$$

with

$$\delta = (\alpha^{(j)} - \alpha^{(i)})\lambda^{(k)} + (\alpha^{(k)} - \alpha^{(j)})\lambda^{(i)} + (\alpha^{(i)} - \alpha^{(k)})\lambda^{(j)}$$

having absolute value

$$|\delta| < c_8 X^{1-\zeta}$$

with

$$c_8 = 6\overline{|\alpha|}.$$

In view of (3.14), (3.15), (3.17) and the last estimate, the right-hand side of equation (3.18) can be estimated from above by

$$c_9 X^{-\zeta} = c_9 \exp(-\zeta \log X) < c_9 \exp(-c_{10} A), \tag{3.19}$$

with

$$c_9 = \frac{|\alpha^{(j)} - \alpha^{(k)}| c_5 + c_8}{c_4(i, j)}, \quad c_{10} = \frac{\zeta}{c_7}.$$

Using the representation (3.16) and applying Theorem 2.1.1 of A. Baker and G. Wüstholz, the left-hand side of (3.18) can be estimated from below, similarly as for Thue equations, by

$$\frac{1}{2} \exp(-C \log A) \quad < \quad \frac{1}{2} \left| \log \left| \frac{(\alpha^{(i)} - \alpha^{(j)}) \mu^{(k)}}{(\alpha^{(i)} - \alpha^{(k)}) \mu^{(j)}} \right| \right.$$

$$\left. + a_1 \log \left| \frac{\eta^{(k)}}{\eta^{(j)}} \right| + \cdots + a_r \log \left| \frac{\eta^{(k)}}{\eta^{(j)}} \right| \right| \tag{3.20}$$

with a large constant C. Comparing (3.19) and (3.20) we get an upper bound A_0 for A.

3.2.3   Reduction, test

The reduction of the bound A_0 is based (similarly as for Thue equations) on the upper estimate

$$\left| \log \left| \frac{\left(\alpha^{(i)} - \alpha^{(j)}\right) \mu^{(k)}}{\left(\alpha^{(i)} - \alpha^{(k)}\right) \mu^{(j)}} \right| + a_1 \log \left| \frac{\eta^{(k)}}{\eta^{(j)}} \right| + \cdots + a_r \log \left| \frac{\eta^{(k)}}{\eta^{(j)}} \right| \right|$$
$$< 2c_9 \exp(-c_{10} A),$$

using Lemma 2.2.2. The possible values of $a_1, \ldots, a_r$ with absolute values under the reduced bound must be tested directly. For each set of exponents we construct β by (3.16). The corresponding solutions $x, y, \lambda = \beta - x - \alpha y$ can be found by solving inequalities. For example if K is totally real and $|y| \geq |x|$, then we must have

$$-|y|^{1-\zeta} < \beta^{(j)} - x - \alpha^{(j)} y < |y|^{1-\zeta}$$

for $j = 1, \ldots, n$, that is

$$\beta^{(j)} - \alpha^{(j)} y - |y|^{1-\zeta} < x < \beta^{(j)} - \alpha^{(j)} y + |y|^{1-\zeta}, \qquad (3.21)$$

which means we only have to consider the values y for which

$$\max_j (\beta^{(j)} - \alpha^{(j)} y - |y|^{1-\zeta}) < \min_j (\beta^{(j)} - \alpha^{(j)} y + |y|^{1-\zeta}).$$

For the possible y we can use (3.21) to find the corresponding values of x. The case $|y| \leq |x|$ is similar. An example with a totally real cubic α and $\zeta = 1/2$ is detailed in [Ga88].

3.2.4   An analogue of the Bilu–Hanrot method

If instead of $\overline{|\lambda|} < X^{1-\zeta}$ we assume the stronger condition $\overline{|\lambda|} < |y|^{1-\zeta}$, then we can use an analogue of the Bilu–Hanrot method. Observe that in our applications in Sections 8.1.1, 8.1.2, 10.1.3 this is sufficient, since in those cases the size of λ is bounded by absolute constants.

Using (3.13) we can bound x by y:

$$|x| \leq |\beta^{(i)}| + |\alpha^{(i)}||y| + |\lambda^{(i)}| < c_{11}|y|.$$

(The positive constants $c_{11}, c_{12}, \ldots$ can be easily calculated.) Similarly as above we have

$$|\beta^{(j)}| > c_{12}|y| \quad \text{for} \quad j \neq i \quad \text{and} \quad |\beta^{(i)}| < c_{13}|y|^{1-n}.$$

The estimates (3.17) and (3.19) remain valid similarly but with $|y|$ instead of X, (3.20) is also valid and we can derive an upper bound for A. If (u_{kj}) is the inverse matrix of $(\log |\eta_k^{(j)}|)$, then (3.16) implies

$$
\begin{aligned}
a_k &= \sum_{j=1}^{r} u_{kj} \log \left| \frac{\beta^{(j)}}{\mu^{(j)}} \right| \\
&= \left(\sum_{j=1}^{r} u_{kj} \right) \log |y| + \sum_{j=1}^{r} u_{kj} \log \left| \frac{\beta^{(j)}}{y(\alpha^{(i)} - \alpha^{(j)})} \right| \\
&\quad + \sum_{j=1}^{r} u_{kj} \log \left| \frac{\alpha^{(i)} - \alpha^{(j)}}{\mu^{(j)}} \right| .
\end{aligned}
\tag{3.22}
$$

We have

$$
\begin{aligned}
\log \left| \frac{\beta^{(j)}}{y(\alpha^{(i)} - \alpha^{(j)})} \right| &= \log \left| \frac{\beta^{(i)} + (\alpha^{(j)} - \alpha^{(i)})y + (\lambda^{(j)} - \lambda^{(i)})}{y(\alpha^{(i)} - \alpha^{(j)})} \right| \\
&= \log \left| \frac{\beta^{(i)} + (\lambda^{(j)} - \lambda^{(i)})}{y(\alpha^{(i)} - \alpha^{(j)})} - 1 \right| \\
&\leq 2 \left| \frac{\beta^{(i)} + (\lambda^{(j)} - \lambda^{(i)})}{y(\alpha^{(i)} - \alpha^{(j)})} \right| \\
&< c_{14}|y|^{-\varsigma} < c_{14} \exp(-c_{15}A).
\end{aligned}
$$

Hence by (3.22) we obtain inequalities of type

$$
|\delta_k \log |y| - a_k + v_k| < c_{16} \exp(-c_{15}A)
$$

with δ_k and v_k that can be easily calculated. Taking any two of these inequalities and eliminating the term with $\log |y|$, we obtain again an inequality of the form

$$
|a_k \vartheta + a_l - \delta| < c_{17} \exp(-c_{15}A) \leq c_{17} \exp(-c_{15} \max(|a_k|, |a_l|))
$$

to which the Davenport lemma (Lemma 2.2.1) can be used. This makes both the reduction and the enumeration processes much faster.

3.3   Relative Thue equations

In our applications, index form equations can often be reduced to relative Thue equations e.g., for sextic fields (Section 8.1), octic fields (Section 10.1), nonic fields (Section 10.2), for relative cubic (Section 9.2) and relative quartic (Section 9.3) extensions.

The enumeration method of Wildanger [Wi97], [Wi00] enabled us to construct a feasible algorithm for solving relative Thue equations (cf. I. Gaál and M. Pohst [GP01]) which we detail here.

Note that formerly B.M.M. de Weger [We95] and N.P. Smart [Sm97] solved some relative Thue equations by using sieve methods, hence their algorithm was not as efficient as ours.

Let $M \subset K$ be algebraic number fields with $m = [M : \mathbb{Q}]$ and $n = [K : M] \geq 3$. The rings of integers of K, M will be denoted by $\mathbb{Z}_K, \mathbb{Z}_M$, respectively. Let $\alpha \in K$ be an integral generator of K over M, $\mu \in M$ an algebraic integer and η an arbitrary unit in M. Consider the relative Thue equation

$$N_{K/M}(X - \alpha Y) = \eta \mu \quad \text{in} \quad X, Y \in \mathbb{Z}_M. \tag{3.23}$$

According to the effective results by A. Baker [Ba90], this equation has only finitely many solutions up to multiplication by units in M. We note that Baker's result was generalized and extended by several authors (for further literature and the latest effective bounds for the sizes of the solutions of (3.23), see Y. Bugeaud and K. Győry [BGy96b]).

3.3.1 Baker's method, reduction

Let $\eta_1, \ldots, \eta_s$ be a system of fundamental units in M. Extend this system to a maximal independent system $\eta_1, \ldots, \eta_s, \varepsilon_1, \ldots, \varepsilon_r$ of units in K. Then any solution $X, Y \in \mathbb{Z}_M$ of (3.23) can be written as

$$X - \alpha Y = \delta(\eta_1)^{b_1} \ldots (\eta_s)^{b_s} (\varepsilon_1)^{a_1} \ldots (\varepsilon_r)^{a_r}. \tag{3.24}$$

Here $\delta \in \mathbb{Z}_K$ is an integral element with relative norm μ. Up to unit factors in K there are only finitely many possibilities for δ, which can be determined using the KASH package [DF97], and the following procedure must be performed for each possible value of δ. Assume for simplicity that the possible roots of unity are also contained in δ.

The exponents $b_1, \ldots, b_s, a_1, \ldots, a_r$ in (3.24) are integers if the above system of independent units is a fundamental system of units. Otherwise, $b_1, \ldots, b_s$, $a_1, \ldots, a_r$ can have a common denominator. In order to make our presentation simpler, we assume that the exponents are integral, otherwise the formulae must be modified in a straightforward way.

Setting

$$X' = \frac{X}{(\eta_1)^{b_1} \ldots (\eta_s)^{b_s}}, \quad Y' = \frac{Y}{(\eta_1)^{b_1} \ldots (\eta_s)^{b_s}}$$

yields

$$X' - \alpha Y' = \delta(\varepsilon_1)^{a_1} \ldots (\varepsilon_r)^{a_r}. \tag{3.25}$$

We will just calculate $a_1, \ldots, a_r$, since the solutions of (3.23) are determined only up to unit factors of M.

For any $\gamma \in K$ we denote by $\gamma^{(11)}, \ldots, \gamma^{(1n)}, \ldots, \gamma^{(m1)}, \ldots, \gamma^{(mn)}$ the conjugates of γ, so that for $1 \leq i \leq m$ the elements $\gamma^{(i1)}, \ldots, \gamma^{(in)}$ are the corresponding relative conjugates of γ over the conjugate field $M^{(i)}$ of M. To simplify our notation, for any i $(1 \leq i \leq m)$ and any distinct j_1, j_2, j_3 $(1 \leq j_1, j_2, j_3 \leq n)$ we introduce a symbol $I = (ij_1 j_2 j_3)$ and set

$$\tau^{(I)} = \tau^{(ij_1 j_2 j_3)} = \frac{\left(\alpha^{(ij_2)} - \alpha^{(ij_3)}\right) \delta^{(ij_1)}}{\left(\alpha^{(ij_1)} - \alpha^{(ij_3)}\right) \delta^{(ij_2)}},$$

$$v_k^{(I)} = v_k^{(ij_1 j_2)} = \frac{\varepsilon_k^{(ij_1)}}{\varepsilon_k^{(ij_2)}} \quad (1 \leq k \leq r),$$

$$\rho^{(I)} = \rho^{(ij_1 j_2)} = \left(v_1^{(ij_1 j_2)}\right)^{a_1} \cdots \left(v_r^{(ij_1 j_2)}\right)^{a_r},$$

and

$$\beta^{(I)} = \beta^{(ij_1 j_2 j_3)} = \frac{\left(\alpha^{(ij_2)} - \alpha^{(ij_3)}\right) \cdot (X' - \alpha Y')^{(ij_1)}}{\left(\alpha^{(ij_1)} - \alpha^{(ij_3)}\right) \cdot (X' - \alpha Y')^{(ij_2)}}.$$

Then we have

$$\beta^{(I)} = \tau^{(I)} \rho^{(I)}.$$

Consider the system of linear equations

$$a_1 \log \left| v_1^{(I)} \right| + \cdots + a_r \log \left| v_r^{(I)} \right| = \log \left| \rho^{(I)} \right| \tag{3.26}$$

in $a_1, \ldots, a_r$ for any $I = (ij_1 j_2 j_3)$ with $1 \leq i \leq m$ and any distinct j_1, j_2, j_3 $(1 \leq j_1, j_2, j_3 \leq n)$ (the equations are independent of j_3). Since $\varepsilon_1, \ldots, \varepsilon_r$ are independent over M, the matrix of coefficients on the left side has rank r. Choosing a set of r linearly independent equations and multiplying by the inverse of the coefficient matrix of the system, we conclude

$$A = \max(|a_1|, \ldots, |a_r|) \leq c_1 \cdot \left| \log \left| \rho^{(I)} \right| \right| \tag{3.27}$$

for a certain set $I = (ij_1 j_2 j_3)$ of indices, where c_1 is the row norm (maximum sum of the absolute values of the elements in a row) of the inverse matrix of the coefficient matrix of (3.26). We choose the set of independent equations so that c_1 becomes as small as possible. Now if $|\rho^{(I)}| < 1$, then (3.27) implies

$$\left| \rho^{(I)} \right| < \exp\left(-\frac{A}{c_1}\right), \tag{3.28}$$

and if $|\rho^{(I)}| > 1$, then the same holds for $|\rho^{(I')}| = 1/|\rho^{(I)}| < 1$ with $I' = (ij_2 j_1 j_3)$. From now on we assume that (3.28) is valid. The following procedure (application of Baker's method, reduction) must be performed for each possible value of i, j_1, j_2 since we cannot predict which of the $|\rho^{(I)}|$ satisfies the crucial inequality (3.28).

Let $1 \leq j_3 \leq n$ be any index distinct from j_1, j_2. Using Siegel's identity we have

$$(\alpha^{(ij_1)} - \alpha^{(ij_2)})(X' - \alpha^{(ij_3)} Y') + (\alpha^{(ij_2)} - \alpha^{(ij_3)})(X' - \alpha^{(ij_1)} Y')$$
$$+ (\alpha^{(ij_3)} - \alpha^{(ij_1)})(X' - \alpha^{(ij_2)} Y') = 0.$$

For $I = (i j_1 j_2 j_3)$ and $I' = (i j_3 j_2 j_1)$ we obtain

$$\beta^{(I)} + \beta^{(I')} = 1. \tag{3.29}$$

Using $|\log x| < 2|x - 1|$ holding for all $|x - 1| < 0.795$ and applying (3.28), from (3.29) we get

$$\left|\log\left(\beta^{(I')}\right)\right| \le 2 \cdot \left|\beta^{(I')} - 1\right| = 2 \cdot \left|\beta^{(I)}\right| \le c_2 \exp\left(-\frac{A}{c_1}\right), \tag{3.30}$$

where $c_2 = 2 \cdot |\tau^{(I)}|$. On the other hand,

$$\left|\log\left(\beta^{(I')}\right)\right| = \left|\log\left(\tau^{(I')}\right) + a_1 \cdot \log\left(v_1^{(I')}\right) + \right.$$
$$\left. \ldots + a_r \cdot \log\left(v_r^{(I')}\right) + a_0 \cdot \log(-1)\right|, \tag{3.31}$$

where $\log$ denotes the principal value of the logarithm and $a_0 \in \mathbb{Z}$ with $|a_0| \le |a_1| + \ldots + |a_r| + 1$. Set $A' = \max(|a_1|, \cdots, |a_r|, |a_0|)$, then $A \le A' \le rA + 1$. Note that (3.30) implies

$$\left|\log\left(\beta^{(I')}\right)\right| \le c_2 \exp\left(-\frac{A' - 1}{rc_1}\right). \tag{3.32}$$

In case the terms in (3.31) are linearly independent, then we can directly apply Theorem 2.1.1 of A. Baker and G. Wüstholz to the linear form in (3.31) to derive a lower bound of type

$$\left|\log\left(\beta^{(I')}\right)\right| \ge \exp(-C \cdot \log A'),$$

which, compared to (3.32), implies an upper bound for A' and A.

Note that if $\log\left(\tau^{(I')}\right)$ in (3.31) is dependent on the other terms, we can reduce the number of variables in the linear form. The variable a_0 can be omitted for totally real fields K.

The bounds obtained by Baker's method are about 10^{20} for $r = 2, 3$ and go up to about 10^{500} for $r = 7, 8$.

Using the estimate (3.32) we can apply Lemma 2.2.2 to reduce the bound. After about 4–5 steps, the reduction procedure stabilizes, i.e., the new bound is not any smaller than the previous bound. Then we stop the procedure. The final reduced bound is usually between 100 and 1000.

3.3.2   Enumeration

Since we usually have $r \ge 3$ for relative Thue equations, in the whole algorithm the most critical step is the test of all possible values of the exponents $a_1, \ldots, a_r$ below the reduced bounds. We use the version of K. Wildanger's enumeration method described in Section 2.3.

Equation (3.29) is just what we had in (2.5). Here the multi-indices $\mathcal{I} \ni I = (ij_1 j_2 j_3)$ range for $1 \le i \le m$ and distinct $1 \le j_1, j_2, j_3 \le n$. The structure of $\beta^{(I)}$ ensures the property (2.6).

We let $I^* = \{I_1, \ldots, I_t\}$ be a set of tuples $I = (ij_1 j_2 j_3)$ with the following properties:

1. If $(ij_1 j_2 j_3) \in I^*$, then either $(ij_2 j_3 j_1) \in I^*$ or $(ij_3 j_2 j_1) \in I^*$.

2. If $(ij_1 j_2 j_3) \in I^*$, then either $(ij_1 j_3 j_2) \in I^*$ or $(ij_3 j_1 j_2) \in I^*$.

3. The vectors

$$
\underline{e}_k = \begin{pmatrix} \log \left| v_k^{(I_1)} \right| \\ \vdots \\ \log \left| v_k^{(I_t)} \right| \end{pmatrix} \qquad (1 \le k \le r)
$$

are linearly independent.

These are tantamount to the conditions in Section 2.3. Since $\varepsilon_1, \ldots, \varepsilon_r$ are multiplicatively independent over M, the last condition can be satisfied if we take sufficiently many tuples. Note that choosing a minimal set of tuples satisfying those conditions reduces the amount of necessary computation considerably.

We are now ready to apply the method in Section 2.3.

3.3.3 An example

Let $M = \mathbb{Q}(\sqrt{2})$ with integral basis $\{1, \omega = \sqrt{2}\}$ and fundamental unit $\eta = 1 + \sqrt{2}$. Consider the equation

$$
X^4 - 2X^3 Y + (-2 - \omega)X^2 Y^2 + (3 + \omega)XY^3 + (1 + \omega)Y^4 = \pm \eta^k
$$

$$
\text{in } X, Y \in \mathbb{Z}_M, \ k \in \mathbb{Z} \ .
$$

The corresponding octic field K is totally real with 7 fundamental units, among them η. Hence we had 6 unknown exponents. The term involving α was independent from the others in the logarithmic linear forms. Baker's method gave a bound of 10^{53}, which was reduced in three steps to 1097, 121, 85, respectively. In the first reduction step we took $H = 10^{350}$ and used a precision of 420 digits. The next steps were much easier, and the whole reduction procedure required about five minutes.

In the final enumeration procedure we had to consider 18 ellipsoids (that is we had to test 18 tuples $(ij_1 j_2 j_3)$). The vector $\underline{g}$ was independent from the vectors $\underline{e}_1, \ldots, \underline{e}_6$ (cf. Section 2.3). This means, that in fact we enumerated quadratic forms in 7 variables, one of them restricted to 1. The reduced bound 85 implied an initial constant $S_0 = 10^{269}$ for the final enumeration. We summarize the enumeration procedure in the following table. In the second and third columns $S > s$ denote the subsequent values $S_k > S_{k+1}$. In the fourth column, Digits, is the precision we used, the fifth column contains the number of tuples found (in the 18 ellipsoids together) and in the last column we display the running time (for the 18 ellipsoids together).

The last line 26. corresponds to the single ellipsoid (2.17). The possible exponents were all tested; if there were corresponding solutions (X, Y) of the equation, this took some seconds. The total CPU time for this example took about 1 hour. The solutions of the equation are

$$(X, Y) = (-\omega, 1 - \omega), (-1 + \omega, -2 + \omega), (\omega, -1), (-1, -1), (0, -1), (1, 0),$$
$$(-1, -1 + \omega), (2 - \omega, -2 + \omega), (-2, -1), (1, -\omega), (1 - \omega, -1),$$
$$(-4 + \omega, -6 + 2\omega), (-1 - \omega, \omega), (\omega, -2 - 2\omega), (-1, 2 - \omega)$$

and of course all multiples of them are units of M.

step	S	s	Digits	tuples	CPUtime
1.	10^{269}	10^{50}	150	0	5 sec
2.	10^{50}	10^{20}	70	0	5 sec
3.	10^{20}	10^{12}	50	0	5 sec
4.	10^{12}	10^{10}	50	0	30 sec
5.	10^{10}	10^{8}	50	4	60 sec
6.	10^{8}	10^{7}	50	42	60 sec
7.	10^{7}	10^{6}	50	195	60 sec
8.	10^{6}	10^{5}	50	2081	180 sec
9.	10^{5}	$10^{4.5}$	50	2185	180 sec
10.	$10^{4.5}$	10^{4}	50	4957	180 sec
11.	10000	6000	50	5005	210 sec
12.	6000	3000	50	7274	240 sec
13.	3000	1500	50	8178	240 sec
14.	1500	1000	50	7306	180 sec
15.	1000	500	50	9113	240 sec
16.	500	250	50	10907	240 sec
17.	250	150	50	10077	240 sec
18.	150	100	50	9265	180 sec
19.	100	50	50	11431	180 sec
20.	50	40	50	6249	120 sec
21.	40	30	50	6297	120 sec
22.	30	20	50	6287	120 sec
23.	20	10	50	7039	120 sec
24.	10	5	50	4459	120 sec
25.	5	3	50	1306	70 sec
26.	3		50	5399	60 sec

3.4   The resolution of norm form equations

There are general criteria on the finiteness of the number of solutions of norm form equations by W.M. Schmidt [Schm72] that were extended to the so-called p-adic case by H.P. Schlickewei [Schl77]. These celebrated results are ineffective. Effective upper bounds for the absolute values of the solutions of norm form equations were given by K. Győry (see e.g., [Gy81]) under some restriction on the coefficients (see below).

Although there is an extensive literature of the explicit resolution of Thue equations (see e.g., Section 3.1), the problem of algorithmic resolution of norm form equations has not yet been investigated. Our purpose is now to fill this gap and to give an efficient method for solving norm form equations under general conditions. The occasion to include this algorithm in this book is given by the fact that we shall use the same tools of Chapter 2 as for the above types of Thue equations.

3.4.1   Preliminaries

Let $\alpha_1 = 1, \alpha_2, \cdots, \alpha_m$ be algebraic integers, linearly independent over $\mathbb{Q}$, let $K = \mathbb{Q}(\alpha_2, \ldots, \alpha_m)$, $L = \mathbb{Q}(\alpha_1, \ldots, \alpha_{m-1})$, and assume that

$$[K : L] \geq 3. \tag{3.33}$$

Let $0 \neq b \in \mathbb{Z}$ and consider the norm form equation

$$N_{K/\mathbb{Q}}(x_1 + \alpha_2 x_2 + \cdots + \alpha_m x_m) = d$$
$$\text{in } x_1, x_2, \ldots, x_m \in \mathbb{Z}, \, x_m \neq 0. \tag{3.34}$$

Using Baker's method, K. Győry gave effective upper bounds for the solutions of norm form equations of the above type, cf. e.g., [Gy81] (see [BGy96b] for improved bounds), reducing the equation to unit equations in two variables. Some of his ideas are used here as well. In order to apply Baker's method it was necessary to make assumptions on the coefficients: (3.33) was the most general assumption of these.

Note that Smart [Sm95] gave a method for solving triangularly connected decomposable form equations, which involve also some special norm form equations, but his purpose was not to consider norm form equations utilizing their special properties, hence his general method is not feasible for norm form equations of the above type.

Our purpose is to give an efficient algorithm (cf. I. Gaál [Ga00c]) for the explicit resolution of equation (3.34). In the course of our method we need to use Baker's method, hence we also have to assume (3.33). In fact we reduce the problem to solving a special type of relative Thue equation over L. One of our goals is to show that by solving equation (3.34) it is sufficient to deal with linear forms in $r(K) - r(L)$ variables, where $r(K)$ resp. $r(L)$ denote the unit rank of K resp. L. The second goal is to show that the enumeration method described in Section

2.3 can be applied in its original form. This becomes an efficient method for the enumeration of small exponents in the corresponding unit equation for reasonable values (up to about 11) of $r(K) - r(L)$.

Let $l = [L : \mathbb{Q}]$, $k = [K : L]$ and denote by $\gamma^{(ij)}$ ($1 \leq i \leq l, 1 \leq j \leq k$) the conjugates of any $\gamma \in K$ so that $\gamma^{(i1)}, \ldots, \gamma^{(ik)}$ are just corresponding relative conjugates of γ over the conjugate field $L^{(i)}$ of L. For elements μ of L we write $\mu^{(i)}$ for $\mu^{(i1)} = \cdots = \mu^{(ik)}$.

Assume that $\eta_1, \ldots, \eta_s$ is a set of fundamental units in L. Let us extend this system to a system of independent units $\eta_1, \cdots, \eta_s, \varepsilon_1, \ldots, \varepsilon_r$ of full rank in K. Denote by q the index of the unit group generated by these units in the whole unit group of K.

Calculate a full set of non-associated integers v of K of norm $\pm d$. The algorithm must be performed for each element v of this set.

Assume that $\underline{x} = (x_1, \ldots, x_m) \in \mathbb{Z}^m$ is a solution of (3.34). Let $l(\underline{x}) = x_1 + \alpha_2 x_2 + \cdots + \alpha_m x_m$. For $1 \leq i \leq l, 1 \leq j \leq k$ we have

$$l^{(ij)}(\underline{x}) = x_1 + \alpha_2^{(i)} x_2 + \ldots + \alpha_{m-1}^{(i)} x_{m-1} + \alpha_m^{(ij)} x_m$$

$$= \zeta^{(ij)} v^{(ij)} \left(\eta_1^{(i)}\right)^{\frac{b_1}{q}} \cdots \left(\eta_s^{(i)}\right)^{\frac{b_s}{q}} \left(\varepsilon_1^{(ij)}\right)^{\frac{a_1}{q}} \cdots \left(\varepsilon_r^{(ij)}\right)^{\frac{a_r}{q}} \quad (3.35)$$

with some integers $b_1, \ldots, b_s, a_1, \ldots, a_r \in \mathbb{Z}$, where ζ is a root of unity and we use throughout a fixed determination of the q-th root of the numbers involved. For any i ($1 \leq i \leq l$) and distinct j_1, j_2, j_3 ($1 \leq j_1, j_2, j_3 \leq k$) we have

$$\left(\alpha_m^{(ij_1)} - \alpha_m^{(ij_2)}\right) l^{(ij_3)}(\underline{x}) + \left(\alpha_m^{(ij_2)} - \alpha_m^{(ij_3)}\right) l^{(ij_1)}(\underline{x})$$

$$+ \left(\alpha_m^{(ij_3)} - \alpha_m^{(ij_1)}\right) l^{(ij_2)}(\underline{x}) = 0,$$

whence

$$\frac{\alpha_m^{(ij_2)} - \alpha_m^{(ij_3)}}{\alpha_m^{(ij_1)} - \alpha_m^{(ij_3)}} \cdot \frac{l^{(ij_1)}(\underline{x})}{l^{(ij_2)}(\underline{x})} + \frac{\alpha_m^{(ij_2)} - \alpha_m^{(ij_1)}}{\alpha_m^{(ij_3)} - \alpha_m^{(ij_1)}} \cdot \frac{l^{(ij_3)}(\underline{x})}{l^{(ij_2)}(\underline{x})} = 1,$$

that is

$$\frac{\left(\alpha_m^{(ij_2)} - \alpha_m^{(ij_3)}\right) \zeta^{(ij_1)} v^{(ij_1)}}{\left(\alpha_m^{(ij_1)} - \alpha_m^{(ij_3)}\right) \zeta^{(ij_2)} v^{(ij_2)}} \left(\frac{\varepsilon_1^{(ij_1)}}{\varepsilon_1^{(ij_2)}}\right)^{\frac{a_1}{q}} \cdots \left(\frac{\varepsilon_r^{(ij_1)}}{\varepsilon_r^{(ij_2)}}\right)^{\frac{a_r}{q}}$$

$$+ \frac{\left(\alpha_m^{(ij_2)} - \alpha_m^{(ij_1)}\right) \zeta^{(ij_3)} v^{(ij_3)}}{\left(\alpha_m^{(ij_3)} - \alpha_m^{(ij_1)}\right) \zeta^{(ij_2)} v^{(ij_2)}} \left(\frac{\varepsilon_1^{(ij_3)}}{\varepsilon_1^{(ij_2)}}\right)^{\frac{a_1}{q}} \cdots \left(\frac{\varepsilon_r^{(ij_3)}}{\varepsilon_r^{(ij_2)}}\right)^{\frac{a_r}{q}} = 1.$$

$$(3.36)$$

This is the unit equation we are going to solve. Note that the denominator q in the exponents will not cause any trouble.

Introduce

$$\gamma^{(ij_1 j_2 j_3)} = \frac{\left(\alpha^{(ij_2)} - \alpha^{(ij_3)}\right) \zeta^{(ij_1)} \nu^{(ij_1)}}{\left(\alpha^{(ij_1)} - \alpha^{(ij_3)}\right) \zeta^{(ij_2)} \nu^{(ij_2)}}, \qquad \rho_k^{(ij_1 j_2)} = \left(\frac{\varepsilon_k^{(ij_1)}}{\varepsilon_k^{(ij_2)}}\right)^{\frac{1}{q}} \qquad (1 \le k \le r)$$

and

$$\tau^{(ij_1 j_2)} = \left(\rho_1^{(ij_1 j_2)}\right)^{a_1} \cdots \left(\rho_r^{(ij_1 j_2)}\right)^{a_r},$$

then we have

$$\beta^{(ij_1 j_2 j_3)} = \frac{\alpha_m^{(ij_2)} - \alpha_m^{(ij_3)}}{\alpha_m^{(ij_1)} - \alpha_m^{(ij_3)}} \cdot \frac{l^{(ij_1)}(x)}{l^{(ij_2)}(x)} = \gamma^{(ij_1 j_2 j_3)} \tau^{(ij_1 j_2)}.$$

for any i $(1 \le i \le l)$ and any distinct j_1, j_2, j_3 $(1 \le j_1, j_2, j_3 \le k)$. Equation (3.36) can be written in the form

$$\beta^{(ij_1 j_2 j_3)} + \beta^{(ij_3 j_2 j_1)} = 1. \tag{3.37}$$

In the following we consider equation (3.37).

3.4.2   Solving the unit equation

By solving the system of linear equations

$$a_1 \log \left|\rho_1^{(ij_1 j_2)}\right| + \cdots + a_r \log \left|\rho_r^{(ij_1 j_2)}\right| = \log \left|\tau^{(ij_1 j_2)}\right| \tag{3.38}$$

in $a_1, \ldots, a_r$ $(1 \le i \le l, \ 1 \le j_1, j_2 \le k, \ j_1 \ne j_2)$, we obtain

$$A = \max(|a_1|, \ldots, |a_r|) \le c_1 \cdot \left|\log \left|\tau^{(ij_1 j_2)}\right|\right| \tag{3.39}$$

for a certain set i, j_1, j_2 of indices, where c_1 (and in the following $c_2, c_3 \ldots$) can be calculated easily (cf. (3.27)). Exchanging j_1 and j_2 if necessary (3.39) implies that there are indices i, j_1, j_2 with

$$\left|\tau^{(ij_1 j_2)}\right| < \exp\left(-\frac{A}{c_1}\right). \tag{3.40}$$

The following steps must be performed for all possible values of i, j_1, j_2.

Let $1 \le j_3 \le k$ be any index distinct from j_1, j_2. Applying (3.40), from (3.37) we get

$$\left|\log\left(\beta^{(ij_3 j_2 j_1)}\right)\right| \le 2 \cdot \left|\beta^{(ij_3 j_2 j_1)} - 1\right|$$
$$= 2 \cdot \left|\beta^{(ij_1 j_2 j_3)}\right| \le c_2 \exp\left(-\frac{A}{c_1}\right). \tag{3.41}$$

On the other hand,

$$\left| \log \left(\beta^{(ij_3 j_2 j_1)} \right) \right| = \left| \log \left(\gamma^{(ij_3 j_2 j_1)} \right) + a_1 \cdot \log \left(\rho_1^{(ij_3 j_2)} \right) \right.$$
$$\left. + \cdots + a_r \cdot \log \left(\rho_r^{(ij_3 j_2)} \right) + a_0 \cdot \log(-1) \right|, \qquad (3.42)$$

where log denotes the principal value of the logarithm, and $a_0 \in \mathbb{Z}$ with $|a_0| \leq |a_1| + \cdots + |a_r| + 1$. Set $A' = \max(|a_1|, \ldots, |a_r|, |a_0|)$, then $A \leq A' \leq rA + 1$. In case the terms in the above linear form are independent (otherwise we can reduce the number of variables), using Theorem 2.1.1 (Section 2.1) of A. Baker and G. Wüstholz and (3.41) we conclude

$$\exp(-C \cdot \log A') \leq \left| \log \left(\beta^{(ij_3 j_2 j_1)} \right) \right| \leq c_2 \exp \left(-\frac{A' + 1}{rc_1} \right), \qquad (3.43)$$

which implies an upper bound A'_B for A' of magnitude 10^{20} up to 10^{500} for $r = 2$ up to 8.

Using (3.42) and (3.43) we have an estimate of type

$$|\xi + a_1 \xi_1 + \cdots + a_r \xi_r + a_0 \xi_0| < c_2 \exp(-c_3 A' - c_4), \qquad (3.44)$$

where

$$\xi = \log \left(\gamma^{(ij_3 j_2 j_1)} \right), \xi_1 = \log \left(\rho_1^{(ij_3 j_2)} \right), \ldots, \xi_r = \log \left(\rho_r^{(ij_3 j_2)} \right), \xi_0 = \log(-1).$$

Using inequality (3.44) we can reduce the bound A'_B for A' by applying Lemma 2.2.2 of Section 2.2.2. The final reduced bound A'_R is usually between 100 and 1000.

Let $\mathcal{I}$ be the set of all tuples $I = (i, j_1, j_2, j_3)$ with $1 \leq i \leq l$, $1 \leq j_1, j_2, j_3 \leq k$ so that j_1, j_2, j_3 are distinct. Introduce

$$\beta^{(I)} = \beta^{(ij_1 j_2 j_3)}, \quad \gamma^{(I)} = \gamma^{(ij_1 j_2 j_3)}, \quad \rho_h^{(I)} = \rho_h^{(ij_1 j_2)} \quad (1 \leq h \leq r).$$

Then we have

$$\beta^{(I)} = \gamma^{(I)} \left(\rho_1^{(I)} \right)^{a_1} \cdots \left(\rho_r^{(I)} \right)^{a_r}$$

and setting $I' = (i, j_3, j_2, j_1)$ equation (3.37) can be written as

$$\beta^{(I)} + \beta^{(I')} = 1.$$

The set $\mathcal{I}$ satisfies (2.6) (cf. Section 2.3).

Let $I^* = \{I_1, \ldots, I_t\}$ be a set of tuples I with the following properties:

1. if $(ij_1 j_2 j_3) \in I^*$, then either $(ij_2 j_3 j_1) \in I^*$ or $(ij_3 j_2 j_1) \in I^*$,
2. if $(ij_1 j_2 j_3) \in I^*$, then either $(ij_1 j_3 j_2) \in I^*$ or $(ij_3 j_1 j_2) \in I^*$,
3. the vectors $\underline{e}_h = \left(\log \left| \rho_h^{(I_1)} \right|, \ldots, \log \left| \rho_h^{(I_t)} \right| \right)$ $(1 \leq h \leq r)$ are linearly independent. These conditions are tantamount to the ones in Section 2.3. We can apply the method of Section 2.3 to enumerate all tuples of exponent vectors satisfying (3.37) with coordinates less than A'_R in absolute value.

3.4.3   Calculating the solutions of the norm form equation

The algorithm of the preceding section gives us all possible tuples $(a_1, \ldots, a_r)$ of exponents in (3.35). For any i, j $(1 \leq i \leq l, 1 \leq j \leq k)$ set

$$\delta^{(i)} = \left(\eta_1^{(i)}\right)^{\frac{b_1}{q}} \cdots \left(\eta_s^{(i)}\right)^{\frac{b_s}{q}},$$

and

$$\gamma^{(ij)} = \zeta^{(ij)} \, \nu^{(ij)} \left(\varepsilon_1^{(ij)}\right)^{\frac{a_1}{q}} \cdots \left(\varepsilon_r^{(ij)}\right)^{\frac{a_r}{q}}.$$

The $\delta^{(i)}$ are not yet known, but the $\gamma^{(ij)}$ are determined by the exponents $(a_1, \ldots, a_r)$. Then we have

$$l^{(ij)}(\underline{x}) = x_1 + \alpha_2^{(i)} x_2 + \cdots + \alpha_{m-1}^{(i)} x_{m-1} + \alpha_m^{(ij)} x_m = \delta^{(i)} \gamma^{(ij)}. \tag{3.45}$$

For any $1 < i \leq l$ we obtain

$$\delta^{(i)}\left(\gamma^{(i1)} - \gamma^{(i2)}\right) = l^{(i1)}(\underline{x}) - l^{(i2)}(\underline{x}) = \left(\alpha_m^{(i1)} - \alpha_m^{(i2)}\right) x_m,$$

hence

$$0 \neq x_m = \delta^{(i)} \, \frac{\gamma^{(i1)} - \gamma^{(i2)}}{\alpha_m^{(i1)} - \alpha_m^{(i2)}},$$

that is

$$\delta^{(i)} = \kappa_i \, \delta^{(1)}, \tag{3.46}$$

with

$$\kappa_1 = 1, \quad \kappa_i = \frac{\alpha_m^{(i1)} - \alpha_m^{(i2)}}{\gamma^{(i1)} - \gamma^{(i2)}} \frac{\gamma^{(11)} - \gamma^{(12)}}{\alpha_m^{(11)} - \alpha_m^{(12)}} \quad \text{for} \quad i = 2, \ldots, l.$$

Substituting our expressions into the original equation (3.34) it can be written in the form

$$\prod_{i=1}^{l} \prod_{j=1}^{k} l^{(ij)}(\underline{x}) = d,$$

whence we obtain

$$\prod_{i=1}^{l} \prod_{j=1}^{k} \left(\kappa_i \, \delta^{(1)} \, \gamma^{(ij)}\right) = d,$$

that is

$$\left(\delta^{(1)}\right)^{kl} = d \left(\prod_{i=1}^{l} \prod_{j=1}^{k} \gamma^{(ij)}\right)^{-1} \left(\prod_{i=1}^{l} \kappa_i\right)^{-k}, \tag{3.47}$$

from which we can calculate the value of $\delta^{(1)}$. This gives at once the value of $\delta^{(i)}$ by (3.46). Finally, solving the system of linear equations (3.45) $(1 \leq i \leq l, 1 \leq j \leq k)$ in $x_1, \ldots, x_m$ we get the solutions of equation (3.34).

3.4.4   Examples

We illustrate our algorithm by two detailed examples. The basic number field data were calculated by using KASH [DF97].

Example 1

Consider first the field K generated by a root ξ of the polynomial

$$f(x) = x^9 - x^8 - 31x^7 + 8x^6 + 200x^5 - 87x^4 - 97x^3 + 27x^2 + 12x - 1.$$

This field is totally real and has integral basis

$$\{1, \xi, \xi^2, \xi^3, \xi^4, \xi^5, \xi^6, \xi^7, \omega_9\}$$

with

$$\omega_9 = \quad (14800 + 24483\xi + 15778\xi^2 + 15468\xi^3$$
$$+19731\xi^4 + 4153\xi^5 + 1420\xi^6 + 4197\xi^7 + \xi^8)/25349.$$

The discriminant of the field is $D_K = 107226034120512 = 2^6 \cdot 3^3 \cdot 37^3 \cdot 107^3$.

The field K has a totally real cubic subfield L generated by α defined by the polynomial $g(x) = x^3 - x^2 - 3x + 1$ with discriminant $D_L = 148 = 2^2 \cdot 37$. (Note that K has also another cubic subfield generated by the root of $x^3 - x^2 - 4x + 1$ with discriminant $321 = 3 \cdot 107$, but this is not interesting for our argument). The field L has integral basis $\{1, \alpha, \alpha^2\}$ and fundamental units

$$\eta_1 = \alpha, \quad \eta_2 = 2\alpha - 1.$$

These elements have the following coefficients in the integral basis of K:

$$\eta_1 = (-430, -703, -454, -472, -568, -117, -42, -122, 736),$$
$$\eta_2 = (-6383, -10561, -6838, -6694, -8428, -1791, -626, -1811, 10936).$$

These units together with

$$\varepsilon_1 = (328, 539, 346, 360, 433, 89, 32, 93, -561),$$
$$\varepsilon_2 = (758, 1242, 800, 832, 1001, 206, 74, 215, -1297),$$
$$\varepsilon_3 = (3590, 5940, 3838, 3746, 4739, 1010, 352, 1018, -6148),$$
$$\varepsilon_4 = (6055, 10022, 6492, 6334, 7995, 1702, 594, 1718, -10375),$$
$$\varepsilon_5 = (103, 164, 108, 112, 135, 28, 10, 29, -175),$$
$$\varepsilon_6 = (6225, 10295, 6682, 6551, 8218, 1745, 611, 1767, -10670)$$

form a system of fundamental units in K. (Hence $s = 2, r = 6, q = 1$.)

Consider the norm form equation

$$N_{K/\mathbb{Q}}(x_1 + \alpha x_2 + \alpha^2 x_3 + \xi x_4) = \pm 1 \tag{3.48}$$
$$\text{in } x_1, x_2, x_3, x_4 \in \mathbb{Z} \quad \text{with} \quad x_4 \neq 0.$$

We had $c_1 = 0.763$ and $c_2 = 4.291$ for all possible i, j_1, j_2. Since our example is a totally real one, we did not have to use a_0. Baker's method gave

$$A = \max(|a_1|, \ldots, |a_6|) \le 10^{36} = A_B.$$

In the reduction procedure we had dimension 7, $c_3 = 1/c_1, c_4 = 0$. The following table summarizes the steps of the reduction procedure. Note that in each step we had to perform 9 reductions.

	$A <$	$|b_1| >$	$H =$	precision	new bound for A	CPU time
Step I	10^{36}	10^{39}	10^{280}	700 digits	429	20 min
Step II	429	9709	10^{30}	100 digits	49	8 sec
Step III	49	1109	10^{22}	60 digits	36	5 sec
Step IV	36	815	10^{21}	60 digits	35	5 sec

Hence our algorithm gave the reduced bound $A_R = 35$.

In the enumeration process we used

$$I^* = \{(i\,123), (i\,231), (i\,312)|i = 1, 2, 3\},$$

that is we had $t = 9$ ellipsoids to consider. The initial bound was $S = 0.4116 \cdot 10^{153}$ that we got using the reduced bound for A. Note that in this example the vector $\underline{g}$ is linearly dependent on $\underline{e}_1, \ldots, \underline{e}_6$. The following table is a summary of the enumeration process.

	S	s	precision	CPU time	tuples found
Step I	10^{153}	10^{20}	100 digits	10 sec	0
Step II	10^{20}	10^{10}	50 digits	5 sec	0
Step III	10^{10}	10^{8}	50 digits	4 sec	0
Step IV	10^{8}	10^{6}	50 digits	4 sec	2
Step V	10^{6}	10^{5}	50 digits	3 sec	8
Step VI	10^{5}	10^{4}	50 digits	3 sec	16
Step VII	10^{4}	10^{3}	50 digits	3 sec	34
Step VIII	10^{3}	10^{2}	50 digits	3 sec	96
Step IX	10^{2}	10	50 digits	3 sec	133
Step X	10	5	50 digits	5 sec	15
Step XI	5	3	50 digits	5 sec	15
Step XII	3		50 digits	3 sec	34

The last line refers to the enumeration of the ellipsoid (2.17) (Section 2.3) with $S = 3$.

We tested all tuples we found in the enumeration process if they are solutions of (3.36). We found 14 solutions of (3.36), the components were all ≤ 2 in absolute value. For these tuples we calculated the corresponding solutions of the equation (3.48). We obtained the following solutions:

x_1	x_2	x_3	x_4
0	0	0	−1
0	0	1	−1
1	−1	−1	1
0	−1	1	1
1	−1	0	−1
−1	0	1	1
0	2	−1	1
−1	−2	0	1

If (x_1, x_2, x_3, x_4) is a solution, then so also is $(-x_1, -x_2, -x_3, -x_4)$, but we list only one of them.

Example 2

Our second example refers to a more complicated situation. Consider the field K generated by a root ξ of the polynomial

$$\begin{aligned} f(x) &= x^{12} - 80x^{10} - 85x^9 + 568x^8 + 184x^7 - 1041x^6 + 40x^5 \\ &\quad + 432x^4 - 19x^3 - 52x^2 - 2x + 1. \end{aligned}$$

This field is totally real and has integral basis

$$\{1, \xi, \xi^2, \xi^3, \xi^4, \xi^5, \xi^6, \xi^7, \xi^8, \omega_{10}, \omega_{11}, \omega_{12}\}$$

with

$$\begin{aligned} \omega_{10} &= (1 + \xi^3 + 2\xi^4 + \xi^6 + \xi^7 + \xi^8 + \xi^9)/3, \\ \omega_{11} &= (2 + \xi + 2\xi^3 + 2\xi^4 + 2\xi^5 + 2\xi^6\xi^{10})/3, \\ \omega_{12} &= (107761264539 + 9245049222\xi + 31097752879\xi^2 + 40137945519\xi^3 \\ &\quad + 34157911107\xi^4 + 93111405784\xi^5 + 51616938926\xi^6 \\ &\quad + 54389034027\xi^7 + 110416671757\xi^8 + 1812369088\xi^9 \\ &\quad + 25415148001\xi^{10} + \xi^{11})/113333753409. \end{aligned}$$

The field K has a totally real quartic subfield L generated by α defined by the polynomial $g(x) = x^4 - 4x^2 + x + 1$. (Note that K has also a cubic subfield generated by the root of $x^3 + 4x^2 - 2x - 1$ but this is not interesting for our argument). The field L has integral basis $\{1, \alpha, \alpha^2, \alpha^3\}$ and fundamental units

$$\eta_1 = \alpha, \quad \eta_2 = 1 - \alpha, \quad \eta_3 = -2\alpha + \alpha^2 + \alpha^3.$$

The units η_1, η_2, η_3 together with $\varepsilon_1, \ldots, \varepsilon_8$ form a system of fundamental units in K, where the coefficients of $\varepsilon_1, \ldots, \varepsilon_8$ in the integral basis of K are the following:

$$\varepsilon_1 = (-10895130684, 3196295645, -6147000968, 2471771060, 4012072493,$$
$$-8357582270, 202752252, -10392673880, -21467491622, -1074736976,$$
$$-15071211644, 22402348184),$$

$$\varepsilon_2 = (761572045, -223421774, 429676837, -172777352, -280445134,$$
$$584196946, -14172204, 726450169, 1500582390, 75124353,$$
$$1053481036, -1565929104),$$

$$\varepsilon_3 = (97534039010, -28613481877, 55028420322, -22127482530, -35916377883,$$
$$74817712333, -1815053823, 93036007507, 192178618710, 9621127196,$$
$$134918633553, -200547525759),$$

$$\varepsilon_4 = (53135222443, -15588237092, 29978737346, -12054752441, -19566755165,$$
$$40759675309, -988817143, 50684755933, 104696306673, 5241459687,$$
$$73501842816, -109255573732),$$

$$\varepsilon_5 = (-137633535407, 40377438576, -77652439063, 31224828557, 50682797994,$$
$$-105577768977, 2561283431, -131286213220, -271189658479,$$
$$-13576693470, -190388183616, 282999302268),$$

$$\varepsilon_6 = (22062864796, -6472564754, 12447804013, -5005387339, -8124529892,$$
$$16924276765, -410577392, 21045379385, 43472114179, 2176364587,$$
$$30519515049, -45365218051),$$

$$\varepsilon_7 = (-62200893641, 18247825609, -35093562605, 14111475601, 22905139427,$$
$$-47713891702, 1157523982, -59332340433, -122559077246, -6135731847,$$
$$-86042366935, 127896224154),$$

$$\varepsilon_8 = (465526096893, -136571013123, 262648465963, -105613596395,$$
$$-171427445547, 357101971563, -8663180146, 444057171072,$$
$$917260919255, 45921258230, 643961282807, -957205380390).$$

Hence $s = 3$, $r = 8$, $q = 1$.

Consider the norm form equation

$$N_{K/\mathbb{Q}}(x_1 + \alpha x_2 + \alpha^2 x_3 + \alpha^3 x_4 + \xi x_5) = \pm 1 \qquad (3.49)$$
$$\text{in } x_1, x_2, x_3, x_4, x_5 \in \mathbb{Z} \text{ with } x_5 \neq 0.$$

We had $c_1 = 0.5187$ and $c_2 = 3.9495$ for all possible i, j_1, j_2. Since our example is a totally real one, we did not have to use a_0. Baker's method gave

$$A = \max(|a_1|, \ldots, |a_8|) \leq 10^{46} = A_B.$$

In the reduction procedure we had dimension 9, $c_3 = 1/c_1$, $c_4 = 0$. The following table summarizes the steps of the reduction procedure. Note that in each step we had to perform 12 reductions.

	$A <$	$|b_1| >$	$H =$	precision	new bound for A	CPU time
Step I	10^{46}	10^{48}	10^{440}	1100 digits	472	98 min
Step II	472	23884	10^{40}	100 digits	45	60 sec
Step III	45	2277	10^{35}	80 digits	40	60 sec
Step IV	40	2024	10^{30}	80 digits	34	60 sec

Hence our algorithm gave the reduced bound $A_R = 34$.
In the enumeration process we used

$$I^* = \{(i123), (i231), (i312)|i = 1, 2, 3, 4\},$$

that is we had $t = 12$ ellipsoids to consider. The initial bound was $S = 0.128 \cdot 10^{174}$ that we got using the reduced bound for A. Note that also in this example the vector g is linearly dependent on $\underline{e}_1, \ldots, \underline{e}_8$. The following table is a summary of the enumeration process.

	S	s	precision	CPU time	tuples found
Step I	10^{174}	10^{20}	100 digits	15 sec	0
Step II	10^{20}	10^{10}	50 digits	10 sec	2
Step III	10^{10}	10^8	50 digits	10 sec	2
Step IV	10^8	10^6	50 digits	8 sec	24
Step V	10^6	10^5	50 digits	5 sec	26
Step VI	10^5	10^4	50 digits	15 sec	91
Step VII	10^4	10^3	50 digits	15 sec	178
Step VIII	1000	500	50 digits	15 sec	57
Step IX	500	250	50 digits	10 sec	45
Step X	250	120	50 digits	10 sec	37
Step XI	120	60	50 digits	12 sec	60
Step XII	60	30	50 digits	10 sec	24
Step XIII	30	15	50 digits	10 sec	17
Step XIV	15	7	50 digits	10 sec	18
Step XV	7	4	50 digits	10 sec	16
Step XVI	4		50 digits	3 sec	125

The last line refers to the enumeration of the ellipsoid (2.17) (Section 2.3) with $S = 4$.

We tested all tuples we found in the enumeration process if they are solutions of (3.36). We found 5 solutions of (3.36), the components were all ≤ 1 in absolute value. For these tuples we calculated the corresponding solutions of the equation

(3.49). We obtained the following solutions:

x_1	x_2	x_3	x_4	x_5
-2	4	0	-1	-1
1	3	0	-1	1
0	1	-1	0	1
0	0	0	0	1

If $(x_1, x_2, x_3, x_4, x_5)$ is a solution, then so also is $(-x_1, -x_2, -x_3, -x_4, -x_5)$, but we list only one of them.

4

Index Form Equations in General

In this chapter we investigate the structure of the index form (1.1). Discovering special properties of the index form, especially factorization properties, makes the resolution of index form equations much easier. A special situation (which otherwise is frequent in numerical examples) is considered in Section 4.4, when the field K is the composite of its subfields. The general results on composite fields have several applications, see e.g., Sections 8.3, 10.2, 10.3.1 and 10.3.3.

4.1   The structure of the index form

Let K be an algebraic number field of degree n with integral basis $\{1, \omega_2, \ldots, \omega_n\}$ and set

$$L^{(i)}(\underline{X}) = X_1 + \omega_2^{(i)} X_2 + \cdots + \omega_n^{(i)} X_n,$$

where $\gamma^{(i)}$ $(1 \leq i \leq n)$ denote the conjugates of any $\gamma \in K$. Let D_K be the discriminant of K. According to (1.1) we have

$$I(X_2, \ldots, X_n) = \pm \frac{1}{\sqrt{|D_K|}} \prod_{1 \leq i < j \leq n} L_{ij}(\underline{X}), \qquad (4.1)$$

where $L_{ij}(\underline{X}) = L^{(i)}(\underline{X}) - L^{(j)}(\underline{X})$, $(1 \leq i < j \leq n)$ that is for any solution $\underline{x} = (x_2, \ldots, x_n) \in \mathbb{Z}^{n-1}$ of the index form equation (1.2) we have

$$\prod_{1 \leq i < j \leq n} L_{ij}^2(\underline{x}) = D_K. \qquad (4.2)$$

Denote by $\overline{K}$ the normal closure of the field K. Equation (4.2) yields

$$L_{ij}(\underline{x}) = \delta_{ij}\eta_{ij}, \tag{4.3}$$

where δ_{ij} is an algebraic integer such that $(N_{\overline{K}/\mathbb{Q}}(\delta_{ij}))^2$ divides D_K (the candidates for δ_{ij} can be explicitly calculated) and η_{ij} is a unit in $\overline{K}$. Up to associated elements there are only finitely many integers in a number field with given norm that can be explicitly calculated by using the algorithm of U. Fincke and M. Pohst [FP85].

Using Siegel's identity

$$L_{ij}(\underline{x}) + L_{jk}(\underline{x}) + L_{ki}(\underline{x}) = 0$$

(where i, j, k are distinct indices) we come to the unit equation

$$\frac{\delta_{ij}}{\delta_{ik}} \cdot \frac{\eta_{ij}}{\eta_{ik}} + \frac{\delta_{jk}}{\delta_{ik}} \cdot \frac{\eta_{jk}}{\eta_{ik}} = 1, \tag{4.4}$$

where the unknowns are the units η_{ij}/η_{ik} and η_{jk}/η_{ik}. It was K. Győry [Gy76] who first used the *connectedness* of the linear forms $L_{ij}(\underline{x})$ to apply Baker's method to index form equations. Namely, for any distinct $1 \leq r, s \leq n$ solving the unit equation (4.4) for $(i, j, k) = (r, s, 2)$ and $(i, j, k) = (2, r, 1)$ we can determine η_{rs}/η_{2r} and η_{2r}/η_{12} hence also η_{rs}/η_{12}. Thus we can express any η_{rs} by η_{12}. Substituting these expressions into equation (4.2), we can determine η_{12}. Then by (4.3) we can calculate the solutions $\underline{x} = (x_2, \ldots, x_n) \in \mathbb{Z}^{n-1}$ of the index form equation by solving a system of linear equations.

Using the above standard methods it is always possible to reduce an index form equation to a system of unit equations in two variables over $\overline{K}$. These unit equations can be in principle solved by using Baker's method and the above described reduction procedure. This direct way was followed by N.P. Smart [Sm93], [Sm95], [Sm96] (see also [Sm98]) who also described how to reduce the number of variables (exponents) by using the Galois structure of the field. K. Győry [Gy98] gave an improvement of this method using unit equations in $K^{(i)}K^{(j)}$ instead of $\overline{K}$. The disadvantage of this attack is that for non–Galois fields, the above extension fields of K have usually a very large unit rank. Writing the unknown units as a power product of the fundamental units, the exponents are to be determined. As we saw in the preceding sections, both the reduction method and especially the enumeration method for the small exponents under the reduced bound are very sensitive for the unit rank: the running time of these procedures grows fast with the unit rank and the enumeration procedure is not feasible any more for unit ranks ≥ 12.

A similar approach based on solving directly the corresponding unit equations was described by M. Klebel [Kl95], who considered relative power integral bases in normal extensions of low degree of imaginary quadratic fields. This problem is as difficult as to solve index form equations in totally real Galois fields of the same degree.

We always try to reduce the index form equation to simpler types of diophantine equations. This can be done in general for lower degree fields. For higher degree number fields with proper subfields we utilize the factorization of the index form, see Section 4.3 below.

4.2 Using resolvents

Let $K = \mathbb{Q}(\xi)$ where ξ is an algebraic integer of degree n. Then we can represent any $\alpha \in \mathbb{Z}_K$ in the form

$$\alpha = \frac{1}{d}(x_1 + x_2\xi + \cdots + x_n\xi^{n-1})$$

with $x_1, \ldots, x_n \in \mathbb{Z}$ and a common denominator $d \in \mathbb{Z}$. In this case the equation $I(\alpha) = 1$ (which is equivalent to the index form equation (1.2)) can be written as

$$\prod_{1 \le i < j \le n} (\alpha^{(i)} - \alpha^{(j)}) = \pm\sqrt{|D_K|},$$

that is

$$\prod_{1 \le i < j \le n} R_{ij}(\underline{x}) = \pm\frac{d^{n(n-1)/2}}{I(\xi)},$$

where

$$R_{ij}(\underline{x}) = \frac{d\left(\alpha^{(i)} - \alpha^{(j)}\right)}{\xi^{(i)} - \xi^{(j)}}$$

$$= x_2 + x_3(\xi^{(i)} + \xi^{(j)}) + x_4\left(\left(\xi^{(i)}\right)^2 + \xi^{(i)}\xi^{(j)} + \left(\xi^{(j)}\right)^2\right) + \cdots$$

$$+ x_n\left(\left(\xi^{(i)}\right)^{n-2} + \left(\xi^{(i)}\right)^{n-3}\xi^{(j)} + \cdots + \left(\xi^{(j)}\right)^{n-2}\right).$$

The advantage of this formula is obvious: if the Galois group of K is doubly transitive, then the coefficients of $R_{ij}(\underline{x})$ are contained in $\mathbb{Q}(\xi^{(i)}\xi^{(j)}, \xi^{(i)} + \xi^{(j)})$ which is a proper subfield of $\mathbb{Q}(\xi^{(i)}, \xi^{(j)})$ (the automorphism interchanging $\xi^{(i)}$ and $\xi^{(j)}$ leaves it fixed) of degree at most $n(n-1)/2$. This reduces the number of unknown exponents of fundamental units in the corresponding unit equation. A similar idea was used by I. Gaál, A. Pethő and M. Pohst [GPP93], I. Gaál and K. Győry [GGy99], K. Győry [Gy00].

4.3 Factorizing the index form when proper subfields exist

Assume that L is a number field of degree r and K is an extension field of L of relative degree s, hence $[K : \mathbb{Q}] = rs$. Denote by $\gamma^{(ij)}$, $j = 1, \ldots, r$ the image of any $\gamma \in K$ under the embeddings of K into the field of complex numbers that leave the conjugate field $L^{(i)}$ elementwise fixed $i = 1, \ldots, s$. Then our equation $I(\alpha) = 1$ can be written in the form

$$\prod_{(1,1) \le (p_1,q_1) < (p_2,q_2) \le (r,s)} \left(\alpha^{(p_1,q_1)} - \alpha^{(p_1,q_1)}\right) = \pm\sqrt{|D_K|},$$

where the pairs of indices are ordered lexicographically. Obviously

$$\prod_{p=1}^{r} \prod_{1 \leq q_1 < q_2 \leq s} \left(\alpha^{(p,q_1)} - \alpha^{(p,q_1)} \right)$$

is a complete norm giving rise to a nontrivial factor of the index of α. Thus, instead of one complicated equation we get a system of equations consisting of simpler equations (of lower degree). Similar ideas were used by I. Gaál [Ga95], [Ga96a], [Ga98a], [Ga01], I. Gaál and M. Pohst [GP00].

4.4   Composite fields

We have a nice factorization of the index form especially when the field K is the composite of its subfields.

4.4.1   Coprime discriminants

Let L be a number field of degree r with integral basis $\{l_1 = 1, l_2, \ldots, l_r\}$ and discriminant D_L. Denote by $I_L(x_2, \ldots, x_r)$ the index form corresponding to the integral basis $\{l_1 = 1, l_2, \ldots, l_r\}$. Similarly, let M be a number field of degree s with integral basis $\{m_1 = 1, m_2, \ldots, m_s\}$ and discriminant D_M. Denote by $I_M(x_2, \ldots, x_s)$ the index form corresponding to the integral basis $\{m_1 = 1, m_2, \ldots, m_s\}$.

Assume, that the discriminants are coprime, that is

$$(D_L, D_M) = 1. \tag{4.5}$$

Denote by $K = LM$ the composite of L and M. As is known (cf. W. Narkiewicz [Nark74]) the discriminant of K is

$$D_K = D_L^s D_M^r \tag{4.6}$$

and an integral basis of K is given by

$$\{l_i m_j : \ 1 \leq i \leq r, 1 \leq j \leq s\}. \tag{4.7}$$

Hence, any integer α of K can be represented in the form

$$\alpha = \sum_{i=1}^{r} \sum_{j=1}^{s} x_{ij} l_i m_j \tag{4.8}$$

with $x_{ij} \in \mathbb{Z}$ $(1 \leq i \leq r, 1 \leq j \leq s)$.

In this section we formulate a general necessary condition for $\alpha \in \mathbb{Z}_K$ to be generator of a power integral basis of K. The following result was proved by I. Gaál [Ga98a]. Applications of it will be given in Sections 10.2 and 10.3.1.

Theorem 4.4.1 *Assume* $(D_L, D_M) = 1$. *If* α *of (4.8) generates a power integral basis in* $K = LM$, *then*

$$N_{M/\mathbb{Q}}\left(I_L\left(\sum_{i=1}^{s} x_{2i} m_i, \ldots, \sum_{i=1}^{s} x_{ri} m_i\right)\right) = \pm 1 \qquad (4.9)$$

and

$$N_{L/\mathbb{Q}}\left(I_M\left(\sum_{i=1}^{r} x_{i2} l_i, \ldots, \sum_{i=1}^{r} x_{is} l_i\right)\right) = \pm 1. \qquad (4.10)$$

Proof. Assume that α generates a power integral basis in K, that is the coefficients x_{ij} satisfy the index form equation corresponding to the basis (4.7). We show that in the present situation, the index form corresponding to the integral basis (4.7) of K factorizes, and two factors imply equations (4.9) and (4.10), respectively. The conjugates of α are given by

$$\alpha^{(p,q)} = \sum_{i=1}^{r} \sum_{j=1}^{s} x_{ij} l_i^{(p)} m_j^{(q)}$$

$(1 \le p \le r, 1 \le q \le s)$. We have

$$\frac{1}{\sqrt{|D_K|}} \prod_{(1,1)\le(p_1,q_1)<(p_2,q_2)\le(r,s)} \left(\alpha^{(p_1,q_1)} - \alpha^{(p_2,q_2)}\right) = \pm 1,$$

where the pairs are ordered lexicographically. A factor of the above index form is obtained by building the symmetric polynomial

$$\prod_{n=1}^{s} \prod_{1\le i<j\le r} \left(\alpha^{(i,n)} - \alpha^{(j,n)}\right)$$

$$= \prod_{n=1}^{s} \prod_{1\le i<j\le r} \left(\sum_{p=1}^{r}\sum_{q=1}^{s}(l_p^{(i)} m_q^{(n)} - l_p^{(j)} m_q^{(n)}) x_{pq}\right)$$

$$= \prod_{n=1}^{s} \prod_{1\le i<j\le r} \left(\sum_{p=1}^{r}\left((l_p^{(i)} - l_p^{(j)})\sum_{q=1}^{s} x_{pq} m_q^{(n)}\right)\right)$$

$$= \left(\sqrt{|D_L|}\right)^{s} N_{M/\mathbb{Q}}\left(I_L\left(\sum_{q=1}^{s} x_{2q} m_q, \ldots, \sum_{q=1}^{s} x_{rq} m_q\right)\right). \qquad (4.11)$$

Similarly,

$$\prod_{n=1}^{r} \prod_{1 \le i < j \le s} \left(\alpha^{(n,i)} - \alpha^{(n,j)} \right)$$

$$= \prod_{n=1}^{r} \prod_{1 \le i < j \le s} \left(\sum_{p=1}^{r} \sum_{q=1}^{s} (l_p^{(n)} m_q^{(i)} - l_p^{(n)} m_q^{(j)}) x_{pq} \right)$$

$$= \prod_{n=1}^{r} \prod_{1 \le i < j \le s} \left(\sum_{q=1}^{s} \left((m_q^{(i)} - m_q^{(j)}) \sum_{p=1}^{r} x_{pq} l_p^{(n)} \right) \right)$$

$$= \left(\sqrt{|D_M|} \right)^r N_{L/\mathbb{Q}} \left(I_M \left(\sum_{p=1}^{r} x_{p2} l_p, \ldots, \sum_{p=1}^{r} x_{ps} l_p \right) \right). \quad (4.12)$$

The factors containing the discriminants D_L, D_M cancel by dividing by $\sqrt{|D_K|}$ because of (4.6). The remaining two polynomials have integer coefficients. Since these two factors, as well as the remaining factor of the index form attain integer values, and their product is equal to ± 1, hence (4.11) and (4.12) imply (4.9) and (4.10), respectively. $\qquad\square$

4.4.2 Non-coprime discriminants

Let $f, g \in \mathbb{Z}[x]$ be distinct monic irreducible polynomials (over $\mathbb{Q}$) of degrees m and n, respectively. Let φ be a root of f and let ψ be a root of g. We assume that the fields $L = \mathbb{Q}(\varphi)$ and $M = \mathbb{Q}(\psi)$ have only $\mathbb{Q}$ as common subfield and the composite field $K = LM$ has degree mn. We also assume that there is an integer q, $(q > 1)$ such that both f and g have a multiple linear factor (at least square) mod q, that is there exist a_f and a_g in $\mathbb{Z}$ such that

$$f(a_f) \equiv f'(a_f) \equiv 0 \pmod{q},$$
$$g(a_g) \equiv g'(a_g) \equiv 0 \pmod{q}. \quad (4.13)$$

Our assumption implies that q divides both the discriminant $d(f)$ of the polynomial f and the discriminant $d(g)$ of g. In our case the fields we consider are composites of subfields, the polynomial orders $\mathbb{Z}[\varphi]$, $\mathbb{Z}[\psi]$ of which have non-coprime discriminants. This is the case in many interesting examples, some of which are given in Sections 8.3 and 10.3.3.

If one of the polynomials e.g., f is cubic, then also conversely, $q | d(f)$ implies that f has the square of a linear factor in its factorization mod q.

Consider the order $\mathcal{O}_f = \mathbb{Z}[\varphi]$ of the field L, the order $\mathcal{O}_g = \mathbb{Z}[\psi]$ of the field M and the composite order $\mathcal{O}_{fg} = \mathcal{O}_f \mathcal{O}_g = \mathbb{Z}[\varphi, \psi]$ in the composite field $K = ML$. Note that the $\mathbb{Z}$-bases of $\mathcal{O}_f, \mathcal{O}_g, \mathcal{O}_{fg}$ are $\{1, \varphi, \ldots, \varphi^{m-1}\}$, $\{1, \psi, \ldots, \psi^{n-1}\}$ and

$$\{1, \varphi, \ldots, \varphi^{m-1}, \psi, \varphi\psi, \ldots, \varphi^{m-1}\psi, \ldots, \psi^{n-1}, \varphi\psi^{n-1}, \ldots, \varphi^{m-1}\psi^{n-1}\},$$

respectively.

The following theorem was proved by I. Gaál, P. Olajos and M. Pohst [GOP01].

Theorem 4.4.2 *Under the above assumptions the index of any primitive element of the order $\mathcal{O}_{fg}$ is divisible by q.*

As a consequence we have:

Theorem 4.4.3 *Under the above assumptions the order $\mathcal{O}_{fg}$ has no power integral bases.*

Note that a similar phenomena (no power integral bases exist) occurs for composite fields also in other cases, cf. Sections 8.1.4, 10.2, 10.3.

Proof of Theorem 4.4.2. Denote the conjugates of $\varphi \in L$ by $\varphi^{(i)}$ $(1 \leq i \leq m)$ and the conjugates of $\psi \in M$ by $\psi^{(j)}$ $(1 \leq j \leq n)$. Denote by $\gamma^{(i,j)}$ the conjugate of any element $\gamma \in K$ under the automorphism mapping φ to $\varphi^{(i)}$ and ψ to $\psi^{(j)}$ $(1 \leq i \leq m, 1 \leq j \leq n)$.

The discriminants of the polynomials f and g are

$$d(f) = \prod_{1 \leq i < j \leq m} (\varphi^{(i)} - \varphi^{(j)})^2,$$

$$d(g) = \prod_{1 \leq i < j \leq n} (\psi^{(i)} - \psi^{(j)})^2. \tag{4.14}$$

These are also the discriminants of the bases $\{1, \varphi, \ldots, \varphi^{m-1}\}$ of the order $\mathcal{O}_f$ and $\{1, \psi, \ldots, \psi^{n-1}\}$ of the order $\mathcal{O}_g$, respectively. The discriminant of the order $\mathcal{O}_{fg}$ is

$$D(\mathcal{O}_{fg}) = d(f)^n \cdot d(g)^m. \tag{4.15}$$

We can represent any element $\alpha \in \mathcal{O}_{fg}$ in the form

$$\alpha = \sum_{i=0}^{m-1} \sum_{j=0}^{n-1} x_{ij} \varphi^i \psi^j \tag{4.16}$$

with $x_{ij} \in \mathbb{Z}$. The index of α corresponding to the order $\mathcal{O}_{fg}$ is

$$I_{\mathcal{O}_{fg}}(\alpha) = \frac{1}{\sqrt{|D(\mathcal{O}_{fg})|}} \prod_{(i_1, j_1) < (i_2, j_2)} \left| \alpha^{(i_1, j_1)} - \alpha^{(i_2, j_2)} \right|,$$

where the pairs of indices are ordered lexicographically. Now we rearrange the factors in the above product. Using (4.14) and (4.15) we have

$$
I_{\mathcal{O}_{fg}}(\alpha) = \prod_{i=1}^{m} \prod_{1 \le j_1 < j_2 \le n} \left| \frac{\alpha^{(i,j_1)} - \alpha^{(i,j_2)}}{\psi^{(j_1)} - \psi^{(j_2)}} \right|
$$

$$
\times \prod_{j=1}^{n} \prod_{1 \le i_1 < i_2 \le m} \left| \frac{\alpha^{(i_1,j)} - \alpha^{(i_2,j)}}{\varphi^{(i_1)} - \varphi^{(i_2)}} \right|
$$

$$
\times \prod_{\substack{(i_1,j_1) < (i_2,j_2) \\ i_1 \ne i_2 \\ j_1 \ne j_2}} \left| \alpha^{(i_1,j_1)} - \alpha^{(i_2,j_2)} \right|. \tag{4.17}
$$

Obviously, the factors that appear in (4.17) are algebraic integers.

For any $1 \le i_1 < i_2 \le m$ and $1 \le j_1 < j_2 \le n$ we have

$$
\left(\alpha^{(i_1,j_1)} - \alpha^{(i_2,j_1)} \right) + \left(\alpha^{(i_2,j_1)} - \alpha^{(i_2,j_2)} \right) + \left(\alpha^{(i_2,j_2)} - \alpha^{(i_1,j_1)} \right) = 0
$$

which implies the equation

$$
\left(\varphi^{(i_1)} - \varphi^{(i_2)} \right) \varepsilon + \left(\psi^{(j_1)} - \psi^{(j_2)} \right) \eta + \rho = 0 \tag{4.18}
$$

with

$$
\varepsilon = \frac{\alpha^{(i_1,j_1)} - \alpha^{(i_2,j_1)}}{\varphi^{(i_1)} - \varphi^{(i_2)}}, \quad \eta = \frac{\alpha^{(i_2,j_1)} - \alpha^{(i_2,j_2)}}{\psi^{(j_1)} - \psi^{(j_2)}}, \quad \rho = \alpha^{(i_2,j_2)} - \alpha^{(i_1,j_1)}.
$$

According to the remark after (4.17) these elements are algebraic integers lying in the $\mathbb{Z}$-order $\mathcal{O} = \mathcal{O}_{i_1,i_2,j_1,j_2} = \mathbb{Z}[\varphi^{(i_1)}, \varphi^{(i_2)}, \psi^{(j_1)}, \psi^{(j_2)}]$.

Let us fix those indices $1 \le i_1 < i_2 \le m$ and $1 \le j_1 < j_2 \le n$ for which $\varphi^{(i_1)} \equiv \varphi^{(i_2)} \pmod{q}$ and also $\psi^{(j_1)} \equiv \psi^{(j_2)} \pmod{q}$. Consider equation (4.18) modulo q.

By our assumptions $\varphi^{(i_1)} - \varphi^{(i_2)} \equiv 0 \pmod{q}$ and $\psi^{(j_1)} - \psi^{(j_2)} \equiv 0 \pmod{q}$, hence by equation (4.18) we get $\rho = \alpha^{(i_2,j_2)} - \alpha^{(i_1,j_1)} \equiv 0 \pmod{q}$. This is one of the algebraic integer factors of $I(\alpha)$, hence it implies $q | I(\alpha)$. $\qquad\square$

5
Cubic Fields

The first non-trivial examples of power integral bases can be found in cubic number fields. In this chapter we briefly summarize the results obtained for computing power integral bases in cubic fields.

This was an easy case, just a routine matter, but this case has initiated an exciting project for about a decade, to extend the computations to higher degree fields.

We also involve here the infinite parametric family of simplest cubic fields in which, thanks to the results on the corresponding parametric Thue equations, we can easily determine the generators of power integral bases.

5.1 Arbitrary cubic fields

For cubic fields the index form equation (1.2) is just a cubic Thue equation

$$I(x_2, x_3) = \pm 1 \quad \text{in} \quad (x_2, x_3) \in \mathbb{Z}, \tag{5.1}$$

which can easily be solved by the methods described in Section 3.1. In fact we only need to use the Davenport lemma (see Lemma 2.2.1). In Table 11.1 we list all power integral bases of the cubic fields with discriminants $-300 \leq D_K \leq 3137$ computed by I. Gaál and N. Schulte [GS89]. This yields about 100 totally real and about 30 complex cubic fields. Note that this table was computed in 1988. Till then only single Thue equations had been solved, this was the first time that the algorithm was completely automated, which enabled us to solve about 130 Thue equations. This computation was later extended by N. Schulte [Schu89], [Schu91].

There are many citations of this simple paper [GS89]. One of the most important applications is given by B. Kovács and A. Pethő [KP91]. They proved that the base number of a generalized number system of an order of a number field must be a generator of a power integral basis of the order. Using the table of [GS89] they calculated number systems in the ring of integers of some cubic number fields.

We should mention here a corresponding result of J.R. Merriman and N.P. Smart [MS93] who calculated cubic integers of discriminant $2^a 3^b$ by solving Thue–Mahler equations.

5.2   Simplest cubic fields

Let $-1 \leq a \in \mathbb{Z}$ and consider a root ϑ of the polynomial

$$f(x) = x^3 - ax^2 - (a+3)x - 1.$$

The fields $K = \mathbb{Q}(\vartheta)$ are called *simplest cubic fields* following D. Shanks [Sh74]. We consider the problem of power integral bases in the polynomial ring $\mathbb{Z}[\vartheta]$ of K. The index form equation corresponding to the basis $\{1, \vartheta, \vartheta^2\}$ of $\mathbb{Z}[\vartheta]$ is

$$I(x, y) = x^3 + 2ax^2y + (a^2 - a - 3)xy^2 + (-a^2 - 3a - 1)y^3 = \pm 1.$$

Substitute $x = x_0 - ay_0$, $y = y_0$, then the above equation takes the shape

$$x_0^3 - ax_0^2y_0 - (a+3)x_0y_0^2 - y_0^3 = \pm 1,$$

that is we arrive at the parametric family of Thue equations corresponding to the polynomial $f(x)$. E. Thomas [Tho90] (for "large" parameters) and M. Mignotte [Mi93] (for "small" parameters) described all solutions of this family of Thue equations.

Lemma 5.2.1 *If* $-1 \leq a \in \mathbb{Z}$, *then all solutions of the equation*

$$x^3 - ax^2y - (a+3)xy^2 - y^3 = \pm 1 \text{ in } x, y \in \mathbb{Z}$$

are

for arbitrary a	$(x, y) =$	$(\pm 1, 0), (0, \pm 1), (\pm 1, \mp 1);$
for $a = -1$ additionally	$(x, y) =$	$(\mp 9, \pm 5), (\pm 1, \pm 1), (\mp 1, \pm 2),$
		$(\pm 2, \mp 1), (\pm 4, \mp 9), (\pm 5, \pm 4);$
for $a = 0$ additionally	$(x, y) =$	$(\mp 3, \pm 2), (\pm 1, \mp 3), (\pm 2, \pm 1);$
for $a = 2$ additionally	$(x, y) =$	$(\pm 7, \pm 2), (\mp 2, \pm 9), (\pm 9, \mp 7).$

This implies:

Theorem 5.2.1 *Up to equivalence all generators of power integral bases in* $\mathbb{Z}[\vartheta]$ *are the following:*

for arbitrary a	$\alpha = \vartheta, \alpha = -a\vartheta + \vartheta^2, \alpha = (1+a)\vartheta - \vartheta^2;$
for $a = -1$ additionally	$\alpha = 9\vartheta + 4\vartheta^2, \alpha = 5\vartheta + 9\vartheta^2, \alpha = -4\vartheta + 5\vartheta^2,$
	$\alpha = -\vartheta + \vartheta^2, \alpha = 2\vartheta + \vartheta^2, \alpha = \vartheta + 2\vartheta^2;$
for $a = 0$ additionally	$\alpha = 2\vartheta + \vartheta^2, \alpha = -3\vartheta + 2\vartheta^2, \alpha = -\vartheta + 3\vartheta^2;$
for $a = 2$ additionally	$\alpha = 3\vartheta + 2\vartheta^2, \alpha = -20\vartheta + 9\vartheta^2, \alpha = -23\vartheta + 7\vartheta^2.$

6
Quartic Fields

In the cubic case the index form equation was a cubic Thue equation. In quartic fields, the index form equation has already degree six and three variables. The resolution of such an equation can yield a difficult problem. The main goal of this Chapter is to point out that in the quartic case the index form equation can be reduced to a cubic and some corresponding quartic Thue equations (see Section 6.1). This means that in fact the index form equations in quartic fields are not much harder to solve than in the cubic case.

There is a series of papers about solving index form equations in quartic fields by I. Gaál, A. Pethő and M. Pohst, cf. [GPP91a]–[GPP96]. We considered quartic fields with a quadratic subfield in [GPP91a] reducing the problem to finding squares (in fact elements of the form $y^2 + a$) in second order linear recurrence sequences. In the cyclic case [GPP91b] the unit equation corresponding to the index form equation is especially simple (see Table 11.2.3). In totally complex quartic fields the general procedure becomes much simpler, see Section 6.4. A very interesting case is the case of biquadratic fields of type $K = \mathbb{Q}(\sqrt{m}, \sqrt{n})$ that we shall consider in Section 6.5.

Interesting tables of the distribution of minimal indices and about the average behaviour of minimal indices can be found in Sections 11.2.1, 11.2.2, respectively.

First we introduce an algorithm, applicable for any quartic fields (Section 6.1). Note that the above mentioned algorithms for special quartic fields are sometimes more efficient in the fields they apply to. Again, we shall see some nice infinite parametric families of fields, like the family of simplest quartic fields (Section 6.2), and some families of totally complex quartic fields (Section 6.4.1).

These methods have also applications. I. Nemes [Ne91] developed further the method of [GPP91a] to solving equations of type $G_n = P(x)$ where G_n is a second order linear recurrence sequence, $P(x)$ is a polynomial of low degree. He applied his procedure to solving index form equations in quartic fields.

6.1   Algorithm for arbitrary quartic fields

Let $K = \mathbb{Q}(\xi)$ be a quartic number field and $f(x) = x^4 + a_1 x^3 + a_2 x^2 + a_3 x + a_4 \in \mathbb{Z}[x]$ the minimal polynomial of ξ. We represent any $\alpha \in \mathbb{Z}_K$ in the form

$$\alpha = \frac{a_\alpha + x\xi + y\xi^2 + z\xi^3}{d} \tag{6.1}$$

with $a_\alpha, x, y, z \in \mathbb{Z}$, and with a common denominator $d \in \mathbb{Z}$.

Our algorithm enables us to solve even the more general equation

$$I(\alpha) = m \quad (\alpha \in \mathbb{Z}_K) \tag{6.2}$$

for $0 < m \in \mathbb{Z}$. Finding the smallest m for which (6.2) is solvable and calculating the corresponding elements α allows us to compute the minimal index of K and all elements of this minimal index.

6.1.1   The resolvent equation

In [GPP93] we proved:

Theorem 6.1.1 *Let $i_m = d^6 m / n$ where n is the index of ξ. The element α of (6.1) is a solution of (6.2) if and only if there is a solution $(u, v) \in \mathbb{Z}^2$ of the cubic equation*

$$\begin{aligned} F(u, v) = \ & u^3 - a_2 u^2 v + (a_1 a_3 - 4a_4)uv^2 \\ & + (4a_2 a_4 - a_3^2 - a_1^2 a_4)v^3 = \pm i_m \end{aligned} \tag{6.3}$$

such that (x, y, z) satisfies

$$\begin{aligned} Q_1(x, y, z) = \ & x^2 - xya_1 + y^2 a_2 + xz(a_1^2 - 2a_2) + yz(a_3 - a_1 a_2) \\ & + z^2(-a_1 a_3 + a_3^2 + a_4) = u \ , \\ Q_2(x, y, z) = \ & y^2 - xz - a_1 yz + z^2 a_2 = v \ . \end{aligned} \tag{6.4}$$

A similar idea was used by A. Bremner [Br85] in a special quartic field. Independently of our method D. Koppenhöfer [Kop94], [Kop95] proved an analogous result using algebraic tools.

Note that literally the same statement is used by F. Leprevost [Le99] and by A. Pethő [Pe01] to so-called index form surfaces.

Proof. We divide equation (6.2) by $I(\xi) = n$ to get

$$\frac{I(\alpha)}{I(\xi)} = \frac{m}{n} . \tag{6.5}$$

Introducing $\alpha' = d \cdot \alpha = a + x\xi + y\xi^2 + z\xi^3$, equation (6.5) becomes

$$\frac{I(\alpha')}{I(\xi)} = \frac{d^6 m}{n} = i_m . \tag{6.6}$$

In the sequel we denote the conjugates of any element $\gamma \in K$ over $\mathbb{Q}$ by γ_i, $i = 1, \ldots, 4$. We can rewrite (6.6) as

$$\prod_{(i,j,k,l)} \left(\frac{\alpha'_i - \alpha'_j}{\xi_i - \xi_j} \frac{\alpha'_k - \alpha'_l}{\xi_k - \xi_l} \right) = \pm i_m, \tag{6.7}$$

where the product is taken for $(i, j, k, l) = (1, 2, 3, 4)$, $(1, 3, 2, 4)$, $(1, 4, 2, 3)$. Let $\theta_{ijkl} = \xi_i \xi_j + \xi_k \xi_l$ and $\theta = \theta_{1234}$. The factor in (6.7) corresponding to the tuple (i, j, k, l) can be expressed as a linear polynomial $u - \theta_{ijkl} v$ of θ_{ijkl}: the coefficients are just $u = Q_1(x, y, z)$, $v = Q_2(x, y, z)$ of (6.4). Then equation (6.7) becomes

$$F(u, v) := (u - \theta_{1234} v)(u - \theta_{1324} v)(u - \theta_{1423} v) = \pm i_m.$$

The coefficients of this equation are obviously symmetric polynomials of ξ_i, $i = 1, \ldots, 4$. Hence, we can express them by the coefficients of $f(t)$ yielding (6.3). □

We note that $F(u, 1)$ is the cubic resolvent of the polynomial f. According to L.C. Kappe and B. Warren [KW89] this polynomial is reducible if the Galois group of K is C_4, V_4 or D_4 (cyclic group of four elements, Klein four group, dihedral group of 8 elements) and irreducible, if the Galois group is S_4 or A_4. In the reducible case equation (6.3) is trivial to solve, otherwise it is a cubic Thue equation that can be easily solved by the methods of Section 3.1.

6.1.2 The quartic Thue equations

We describe now how to solve the system of equations (6.4) (cf. [GPP96]).

Set $d_1 = \gcd(x, y, z)$. Then it follows from (6.3) and (6.4) that d_1^2 divides $\gcd(u, v)$ and d_1^6 divides i_m, hence there remain very few possible values for d_1. The following steps must be performed for all these values. Let

$$x_1 = \frac{x}{d_1}, \quad y_1 = \frac{y}{d_1}, \quad z_1 = \frac{z}{d_1}. \tag{6.8}$$

Then $\gcd(x_1, y_1, z_1) = 1$ and from (6.4) we obtain

$$Q_1(x_1, y_1, z_1) = \frac{u}{d_1^2}, \quad Q_2(x_1, y_1, z_1) = \frac{v}{d_1^2}. \tag{6.9}$$

Using $u_1 = u/d_1^2$, $u_2 = v/d_1^2$, introduce the form

$$Q_0(X, Y, Z) = u_1 Q_2(X, Y, Z) - u_2 Q_1(X, Y, Z).$$

By equation (6.9) we have

$$Q_0(x, y, z) = 0. \tag{6.10}$$

Rewriting $Q_0(X, Y, Z)$ as a sum of squares and applying Theorems 3 and 5 of Chapter 7 of L.J. Mordell [Mo69], we can easily decide if (6.10) has a non-trivial solution. If yes, let (x_Q, y_Q, z_Q) be a non-trivial solution of equation (6.10). Without loss of generality we may assume $z_Q \neq 0$.

Following [Mo69] we write all solutions x, y, z of (6.10) in a parametric form

$$
\begin{aligned}
x &= r x_Q + p, \\
y &= r y_Q + q, \\
z &= r z_Q
\end{aligned}
\tag{6.11}
$$

with rational parameters r, p, q and substitute them into (6.10). Since (x_Q, y_Q, z_Q) is a solution of (6.10), we get

$$r(c_1 p + c_2 q) = c_3 p^2 + c_4 pq + c_5 q^2, \tag{6.12}$$

where the integer parameters $c_1, \ldots, c_5$ are easily calculated. Next we multiply (6.11) by $(c_1 p + c_2 q)$. In addition, we multiply these equations by the square of the common denominator of p, q to obtain all integer relations. We divide those by $\gcd(p, q)^2$ and get

$$
\begin{aligned}
k \cdot x &= f_x(p, q) = c_{11} p^2 + c_{12} pq + c_{13} q^2, \\
k \cdot y &= f_y(p, q) = c_{21} p^2 + c_{22} pq + c_{23} q^2, \\
k \cdot z &= f_z(p, q) = c_{31} p^2 + c_{32} pq + c_{33} q^2,
\end{aligned}
\tag{6.13}
$$

where $k \in \mathbb{Z}^{\geq 0}$, $c_{ij} \in \mathbb{Z}$ ($1 \leq i, j \leq 3$) and the parameters p, q are coprime. Substituting the representation (6.13) for x, y, z into the equations of (6.9) we get

$$
\begin{aligned}
F_i(p, q) &= Q_i(f_x(p, q), f_y(p, q), f_z(p, q)) = k^2 u_i \\
&\quad (p, q \in \mathbb{Z}; \ i = 1, 2).
\end{aligned}
\tag{6.14}
$$

To characterize the potential values of k we use the following lemma (cf. [GPP96]).

Lemma 6.1.1 *We have*

$$|z_Q|^{-3} |\det(C)| = 4|\det(Q_0)| = |F(u, v)| = i_m \neq 0,$$

where $C = (c_{ij})$ is the above 3 by 3 matrix and Q_0 is the matrix of the quadratic form $Q_0(X, Y, Z)$.

Proof. Straightforward calculation.   □

The above lemma implies that C is a regular matrix. Let

$$C^{-1} = \left(\frac{\bar{c}_{ij}}{\det(C)} \right)_{1 \le i, j \le 3}$$

with integers $\bar{c}_{ij}$. If we consider (6.13) as a system of linear equations in the variables p^2, pq, q^2 we obtain by Cramer's rule

$$
\begin{aligned}
\det(C) \cdot p^2 &= k \cdot (\bar{c}_{11}x + \bar{c}_{12}y + \bar{c}_{13}z), \\
\det(C) \cdot q^2 &= k \cdot (\bar{c}_{31}x + \bar{c}_{32}y + \bar{c}_{33}z).
\end{aligned}
$$

This implies that $k \ne 0$. The number k divides both left-hand sides and—because of p, q being coprime integers—k divides

$$\frac{\det(C)}{\gcd\{c_{ij} | 1 \le i, j \le 3\}^3}.$$

This restricts the potential values of k. We note that according to Lemma 6.1.1 the best choice for z_Q is the one with the smallest number of divisors among the non-zero coordinates of the known non-trivial integer solutions of (6.10).

To diminish the number of possibilities for k further, we check whether the system (6.13) is solvable modulo k so that the gcd of the residue classes of p and q is coprime to k.

For all potential values of k we need to solve the equations

$$F_1(p, q) = Q_1(f_x(p, q), f_y(p, q), f_z(p, q)) = w_1 := \frac{k^2 u}{d_1^2}, \quad (6.15)$$

$$F_2(p, q) = Q_2(f_x(p, q), f_y(p, q), f_z(p, q)) = w_2 := \frac{k^2 v}{d_1^2}. \quad (6.16)$$

The main result of [GPP96] is that at least one of these equations is a quartic Thue equation splitting in the original field K.

Theorem 6.1.2 *If $v = 0$, then equation (6.15), in case $v \ne 0$ equation (6.16), is a quartic Thue equation over K. Hence, the resolution of the system of equations (6.4) can be reduced to the resolution of some quartic Thue equations splitting in the original field K.*

As a consequence (see [GPP96]) we have:

Corollary 6.1.1 *In any quartic field all power integral bases can be determined by solving a cubic and some quartic Thue equations.*

For all solutions p, q of (6.14) we calculate the corresponding values of x, y, z from (6.13) and test whether (6.2) holds.

6.1.3   Proof of the theorem on the quartic Thue equations

We now prove Theorem 6.1.2.

We substitute $X = Y - a_1$ in the minimal polynomial f of ξ and obtain another quartic form via

$$g(Y, Z) = Z^4 f(Y - a_1) = N_{K/\mathbb{Q}}(Y - a_1 Z - \xi Z). \qquad (6.17)$$

We begin with two preparatory lemmata and then consider the cases $v = 0$ and $v \neq 0$ separately.

Lemma 6.1.2 *For $(x_Q, y_Q, z_Q) \neq (0, 0, 0)$ we have*

$$|Q_1(x_Q, y_Q, z_Q)| + |Q_2(x_Q, y_Q, z_Q)| > 0.$$

Proof. Let us assume that $|Q_1(x_Q, y_Q, z_Q)| + |Q_2(x_Q, y_Q, z_Q)| = 0$. If z_Q is zero, then $Q_2(x_Q, y_Q, z_Q) = 0$ implies $y_Q = 0$ and from $Q_1(x_Q, y_Q, z_Q) = 0$ we get $x_Q = 0$, clearly a contradiction. For $z_Q \neq 0$ the equation $Q_2(x_Q, y_Q, z_Q) = 0$ yields

$$x_Q = \frac{y_Q^2 - a_1 y_Q z_Q + a_2 z_Q^2}{z_Q}. \qquad (6.18)$$

Substituting this expression for x_Q into $Q_1(x_Q, y_Q, z_Q) = 0$ we obtain

$$y_Q^4 - 3a_1 y_Q^3 z_Q + (a_2 + 3a_1^2) y_Q^2 z_Q^2 + (a_3 - 2a_1 a_2 - a_1^3) y_Q z_Q^3$$
$$+ (a_1^2 a_2 + a_4 - a_1 a_3) z_Q^4 = 0 ,$$

hence,

$$f\left(\frac{y_Q}{z_Q} - a_1\right) = 0.$$

But this is again a contradiction since f has no rational roots.   □

Lemma 6.1.3 *All solutions (u, v) of (6.3) and $(x_Q, y_Q, z_Q) \neq (0, 0, 0)$ of (6.10) satisfy:*
(i) $y_Q = 0$ and $z_Q = 0$ imply $v = 0$;
(ii) $v \neq 0$ has $Q_2(x_Q, y_Q, z_Q) \neq 0$ as a consequence.

Proof. (i)The conditions $y_Q = 0$ and $z_Q = 0$ yield $Q_2(x_Q, y_Q, z_Q) = 0$ and therefore $Q_1(x_Q, y_Q, z_Q) \neq 0$ because of Lemma 6.1.2. From (6.10) we obtain

$$u Q_2(x_Q, y_Q, z_Q) = v Q_1(x_Q, y_Q, z_Q),$$

hence, $v = 0$.
(ii) This is immediate from $u Q_2(x_Q, y_Q, z_Q) = v Q_1(x_Q, y_Q, z_Q)$ and Lemma 6.1.2.   □

The case $v = 0$. We show that if $v = 0$ in (6.4), then equation (6.15) is a quartic Thue equation over K.

Using substitution (6.11) and $v = 0$ we get

$$f_y(p, q) - a_1 f_z(p, q) - \xi f_z(p, q) = -uq(z_Q p - (y_Q + z_Q \xi)q).$$

We put

$$h(p, q) = z_Q p - (y_Q + z_Q \xi)q,$$

$$H(p, q) = N_{K/\mathbb{Q}}(h(p, q)),$$

and express the coefficients of $H(p, q)$ by $y_Q, z_Q, a_1, \ldots, a_4$.

On the other hand we consider

$$F_1(p, q) = Q_1(f_x(p, q), f_y(p, q), f_z(p, q)).$$

For $v = 0$ equation (6.10) simply becomes $Q_2(x_Q, y_Q, z_Q) = 0$ because of $(u, v) \neq (0, 0)$. Expressing x_Q as in (6.18) and substituting this into $F_1(p, q)$ we obtain

$$\frac{z_Q^2}{u^2} F_1(p, q) = H(p, q).$$

Clearly one root of $H(p, 1) = 0$ is $\xi + y_Q/z_Q$ which is a primitive element of K. Hence, the form $F_1(p, q)$ is irreducible and has a root in K, i.e. (6.15) is a Thue equation over K.

The case $v \neq 0$. We show that if $v \neq 0$ in (6.4), then equation (6.16) is a quartic Thue equation over K.

I. We first prove that $F_2(p, q)$ has a root in K.

Again we consider

$$f_y(p, q) - a_1 f_z(p, q) - \xi f_z(p, q) = t_2 p^2 + t_1 pq + t_0 q^2 , \qquad (6.19)$$

where the coefficients on the right-hand side are

$$
\begin{aligned}
t_2 &= v(y_Q - z_Q a_1 - z_Q \xi) , \\
t_1 &= -2vx_Q + 2va_2 z_Q - uz_Q + va_1 z_Q \xi , \\
t_0 &= va_1 x_Q - va_2 y_Q + uy_Q - va_3 z_Q + uz_Q \xi - va_2 z_Q \xi .
\end{aligned}
$$

We note that $t_2 \neq 0$. It is well known that $t_2 p^2 + t_1 pq + t_0 q^2$ factorizes over K if and only if its discriminant $D = t_1^2 - 4t_2 t_0$ is a square in K. Adding $4v Q_0(x_Q, y_Q, z_Q) = 0$ to D yields

$$D = z_Q^2((v^2 a_1^2 - 4v^2 a_2 + 4uv)\xi^2 + (2uva_1 - 4v^2 a_3)\xi + u^2 - 4v^2 a_4) ,$$

and adding $4v^2 z_Q^2 f(\xi) = 0$ we indeed obtain

$$D = (z_Q(2v\xi^2 + va_1\xi + u))^2.$$

We conclude that $t_2 p^2 + t_1 pq + t_0 q^2$ factorizes over K in the form

$$t_2 p^2 + t_1 pq + t_0 q^2 = t_2(p - \rho_1 q)(p - \rho_2 q) \tag{6.20}$$

with

$$\rho_1 = \frac{-t_1 - \sqrt{D}}{2t_2}, \quad \rho_2 = \frac{-t_1 + \sqrt{D}}{2t_2}.$$

Similarly to case $v = 0$ we introduce

$$h(p, q) := \frac{1}{v} t_2(p - \rho_1 q) = \frac{1}{v}\left(t_2 p + \frac{t_1 + \sqrt{D}}{2} q\right)$$

$$= z_Q q \xi^2 + (-z_Q p + z_Q a_1 q)\xi + y_Q p - z_Q a_1 p - x_Q q + z_Q a_2 q$$

and its norm

$$H(p, q) = N_{K/\mathbb{Q}}(h(p, q)),$$

whose coefficients can be expressed in terms of $x_Q, y_Q, z_Q, a_1, \ldots, a_4$.

On the other hand we consider $F_2(p, q)$. From Lemma 6.1.3 (ii) we know that $Q_2(x_Q, y_Q, z_Q) \neq 0$, hence, (6.10) yields

$$v = u \frac{Q_1(x_Q, y_Q, z_Q)}{Q_2(x_Q, y_Q, z_Q)}.$$

Substituting this into $F_2(p, q)$ we obtain $H(p, q)$ upon multiplication by $Q_2(x_Q, y_Q, z_Q)$ and division by v^2. Therefore $F_2(p, q)$ has a root in K.

II. We additionally show that the root ρ_1 of $H(p, 1)$ is a primitive element of K implying that (6.16) is a Thue equation over K.

Set

$$\alpha = z_Q \xi^2 + z_Q a_1 \xi + s \quad \text{with} \quad s = z_Q a_2 - x_Q,$$
$$\beta = -z_Q \xi + t \quad \text{with} \quad t = y_Q - z_Q a_1.$$

The element $\rho_1 = -\alpha/\beta \in K$ is certainly not rational since ξ is an irrationality of degree four. Let us assume that ρ_1 is quadratic.

Denote by ξ_i, α_i, β_i $(i = 1, \ldots, 4)$ the corresponding conjugates of ξ, α, β, respectively. Let G be the Galois group of K. If K has a quadratic subfield, then the possibilities for G are V_4, C_4 and D_4 (the Klein four group, the cyclic group of four elements, the dihedral group with 8 elements). If α/β is quadratic, then we may assume that $\alpha_1/\beta_1 = \alpha_3/\beta_3 \neq \alpha_2/\beta_2 = \alpha_4/\beta_4$. The group G must induce an operation on $\{\{\xi_1, \xi_3\}, \{\xi_2, \xi_4\}\}$. From $\alpha_1/\beta_1 = \alpha_3/\beta_3$ we deduce

$$z_Q \xi_1 \xi_3 - t(\xi_1 + \xi_3) - (ta_1 + s) = 0. \tag{6.21}$$

Multiplying by z_Q and substituting $z_Q \xi_i = t - \beta_i$ yields

$$\beta_1 \beta_3 = t^2 + t z_Q a_1 + z_Q s. \tag{6.22}$$

Using (6.21) and (6.22) we get

$$\alpha_1\beta_3 = \xi_1(-z_Qt\xi_3 - z_Qs) - a_1z_Q^2\xi_1\xi_3 - sz_Q\xi_3 + st$$

$$= (t^2 + dz_Qa_1 + z_Qs)(-\xi_1 - \xi_3 - a_1) = \beta_1\beta_3(\xi_2 + \xi_4) \ ,$$

whence

$$\frac{\alpha_1}{\beta_1} = \frac{\alpha_3}{\beta_3} = \xi_2 + \xi_4 \ \text{and} \ \frac{\alpha_2}{\beta_2} = \frac{\alpha_4}{\beta_4} = \xi_1 + \xi_3, \tag{6.23}$$

where the second part follows from the operation of the Galois group. Similarly, (6.21) also holds for $\{\xi_2, \xi_4\}$ instead of $\{\xi_1, \xi_3\}$, which yields $z_Q(\xi_1\xi_3 - \xi_2\xi_4) = t(\xi_1 + \xi_3 - \xi_2 - \xi_4)$ or

$$z_Q(\xi_1\xi_3 - \xi_2\xi_4 + a_1(\xi_1 + \xi_3 - \xi_2 - \xi_4)) = y_Q(\xi_1 + \xi_3 - \xi_2 - \xi_4) \ . \tag{6.24}$$

Because of Lemma 6.1.3 (ii), $v \neq 0$ implies $Q_2(x_Q, y_Q, z_Q) \neq 0$. Now $u - v(\xi_1\xi_3 + \xi_2\xi_4)$ is a factor of $F(u, v)$ (cf. proof of Theorem 6.1.1). We show that $Q_1(x_Q, y_Q, z_Q)/Q_2(x_Q, y_Q, z_Q) = \xi_1\xi_3 + \xi_2\xi_4$, contradicting $F(u, v) = \pm i_m \neq 0$. Since $\xi_1\xi_3 + \xi_2\xi_4 = a_2 - (\alpha_1\alpha_2)/(\beta_1\beta_2)$, it suffices to show (cf. (6.22)) that

$$Q_1(x_Q, y_Q, z_Q) - a_2 Q_2(x_Q, y_Q, z_Q) = -\alpha_1\frac{\alpha_2}{\beta_2}\beta_3. \tag{6.25}$$

Let us start with

$$-\alpha_1\frac{\alpha_2}{\beta_2}\beta_3 = a_1\alpha_1\beta_3 + \alpha_1\alpha_3$$

$$= s^2 + a_1ts + z_Q[z_Q\xi_1^2\xi_3^2 + a_1t\xi_1\xi_3 + (s + a_1t)(\xi_1^2 + \xi_3^2 + a_1(\xi_1 + \xi_3))]$$

where we used (6.23) in the first and (6.21) in the last step. Since by (6.22) the above expression is equal to its conjugate, it belongs to $\mathbb{Q}$, hence it is invariant if we substitute $\{\xi_2, \xi_4\}$ for $\{\xi_1, \xi_3\}$. Therefore the above equals

$$s^2 + a_1ts + \frac{z_Q}{2}[z_Q(\xi_1^2\xi_3^2 + \xi_2^2\xi_4^2) + a_1t(\xi_1\xi_3 + \xi_2\xi_4) - 2a_2(s + a_1t)] \ .$$

Substituting t and using (6.24) we arrive at

$$(s - a_2z_Q)(s + a_1t)$$
$$+\frac{z_Q}{2}\left[z_Q(\xi_1^2\xi_3^2 + \xi_2^2\xi_4^2 - a_1^2(\xi_1\xi_3 + \xi_2\xi_4))\right.$$
$$\left.+y_Q(2a_3 - (\xi_1 + \xi_3 - \xi_2 - \xi_4)(\xi_1\xi_3 - \xi_2\xi_4))\right]$$
$$\doteq -x_Q(s + a_1t) + a_3y_Qz_Q + z_Q^2(a_4 - a_1a_3)$$
$$= Q_1(x_Q, y_Q, z_Q) - a_2Q_2(x_Q, y_Q, z_Q),$$

which proves (6.25).

Remark The above statements can be proved also by direct calculations using a computer algebra system, e.g., Maple [CG88].

6.1.4 Examples

Finally we give some explicit examples for the application of this general method, displaying among others the Thue equations that arise. The examples correspond to totally real quartic fields with Galois groups S_4 or A_4 that can not be dealt with by the special methods. We list the discriminant D_K of the field $K = \mathbb{Q}(\xi)$, the minimal polynomial $f(x)$ of ξ, the Galois group of K, an integral basis $\{\omega_1 = 1, \omega_2, \omega_3, \omega_4\}$ of K, the index n of ξ, and the field index $m(K)$ of K. All m ($1 \le m \le n$) divisible by $m(K)$ are candidates for the minimal index of K. We consider them in increasing order until we obtain elements of index m. For each m we present the cubic equation (6.3), and all corresponding quartic equations ((6.15) or (6.16)) with all occurring right-hand sides. Finally, we present the minimal index $\mu(K)$ of the field K and a complete list of the solutions of the index form equation corresponding to the given integral basis. That is, all generators of power integral bases are of the form $\alpha = a \pm (x_0\omega_2 + y_0\omega_3 + z_0\omega_4)$ where $a \in \mathbb{Z}$ is arbitrary and (x_0, y_0, z_0) is listed.

EXAMPLE 1

$D_K = 1957$, $f(x) = x^4 - 4x^2 - x + 1$, Galois group S_4,
integral basis $\{1, \xi, \xi^2, \xi^3\}$, $n = 1$, $m(K) = 1$,
$m = 1$, $F(u, v) = u^3 + 4u^2v - 4uv^2 - 17v^3 = \pm 1$
$\quad F_1(p, q) = p^4 - 4p^2q^2 - pq^3 + q^4 = \pm 1$
$\quad F_2(p, q) = p^4 + 8p^3q + 18p^2q^2 + 7pq^3 - 3q^4 = \pm 1$
$\quad F_2(p, q) = p^4 + 15p^3q + 76p^2q^2 + 154pq^3 + 101q^4 = \pm 1$
$\quad F_2(p, q) = p^4 - p^3q - 12p^2q^2 + 6pq^3 + 37q^4 = \pm 1$
Minimal index: $\mu(K) = 1$
Solutions: $(-12,1,3),(-8,1,2),(-5,0,1),(-4,0,1),(-4,1,1),(-3,0,1),(0,1,0),$
$(0,2,1),(1,0,0),(1,2,-1),(2,1,-1),(3,1,-1),(4,1,-1),(4,9,-5),(4,33,16),(8,1,-2),$
$(14,3,-4)$

EXAMPLE 2

$D_K = 2777$, $f(x) = x^4 - x^3 - 4x^2 + x + 2$, Galois group S_4,
integral basis $\{1, \xi, \xi^2, \xi^3\}$, $n = 1$, $m(K) = 1$,
$m = 1$, $F(u, v) = u^3 + 4u^2v - 9uv^2 - 35v^3 = \pm 1$,
$\quad F_1(p, q) = p^4 + 3p^3q - p^2q^2 - 6pq^3 - q^4 = \pm 1$
$\quad F_2(p, q) = p^4 + 8p^3q + 19p^2q^2 + 13pq^3 + 2q^4 = \pm 1$
$\quad F_2(p, q) = -p^4 + 3p^3q + 9p^2q^2 - 38pq^3 + 31q^4 = \pm 1$
$\quad F_2(p, q) = 2p^4 + 47p^3q + 393p^2q^2 + 1372pq^3 + 1697q^4 = \pm 1$
$\quad F_2(p, q) = 8(71p^4 + 185p^3q + 144p^2q^2 + 31pq^3 + 2q^4) = \pm 8$
Minimal index: $\mu(K) = 1$
Solutions: $(-6,2,1),(-4,0,1),(-1,1,0),(-1,1,1),(0,-4,1),(0,1,0),(1,0,0),$
$(1,2,-1),(1,10,-4),(2,2,-1),(3,1,-1),(3,2,-1),(5,1,-1),(6,3,-2),(21,1,-5)$

EXAMPLE 3
$D_K = 15188$, $f(x) = x^4 - x^3 - 7x^2 + x + 2$, Galois group S_4,
integral basis $\{1, \xi, \xi^2, (\xi + \xi^3)/2\}$, $n = 2$, $m(K) = 1$,
$m = 1$, $F(u, v) = u^3 + 7u^2v - 9uv^2 - 59v^3 = \pm 32$
  $F_2(p, q) = p^4 + 14p^3q + 52p^2q^2 + 22pq^3 - q^4 = \pm 2, \pm 8, \pm 32$
  $F_2(p, q) = 2(p^4 + 11p^3q + 33p^2q^2 + 19pq^3 - 26q^4) = \pm 2, \pm 8$
  $F_2(p, q) = 2(11p^4 + 119p^3q + 447p^2q^2 + 649pq^3 + 248q^4) = \pm 2, \pm 8$
Minimal index: $\mu(K) = 1$
Solutions: $(1,-2,1),(2,1,-1),(4,1,-1),(12,1,-3)$
We observe, that $I(\xi) = 2$ but the minimal index is 1 and the elements given by
the solutions generate power integral bases. Hence, we found better generators for
K over $\mathbb{Q}$, e.g., a root of the polynomial $x^4 - 7x^3 + 6x + 2$ obtained from the
solution $(2, 1, -1)$.

EXAMPLE 4
$D_K = 157609$, $f(x) = x^4 - 13x^2 - 2x + 19$, Galois group A_4,
integral basis $\{1, \xi, (1 + \xi + \xi^2)/2, (1 + \xi^3)/2\}$, $n = 4$, $m(K) = 2$,
$m = 2$, $F(u, v) = u^3 + 13u^2v - 76uv^2 - 992v^3 = \pm 32$
  $F_2(p, q) = p^4 + 26p^3q + 207p^2q^2 + 490pq^3 + 309q^4 = \pm 2$
  $m = 4$, $F(u, v) = u^3 + 13u^2v - 76uv^2 - 992v^3 = \pm 64$
  $F_1(p, q) = p^4 - 13p^2q^2 - 2pq^3 + 19q^4 = \pm 1, \pm 4$
  $F_2(p, q) = p^4 + 26p^3q + 188p^2q^2 + 96pq^3 - 1792q^4 = \pm 1, \pm 4, \pm 16, \pm 64, \pm 256$
  $F_2(p, q) = 11p^4 + 100p^3q + 262p^2q^2 + 172pq^3 - 81q^4 = \pm 1, \pm 4, \pm 16$
Minimal index: $\mu(K) = 4$
Solutions: $(-6,1,1),(-1,1,0),(0,-3,1),(0,1,0),(1,0,0), (5,1,-1),(24,7,-5)$

EXAMPLE 5
$D_K = 848241$, $f(x) = x^4 - x^3 - 16x^2 - 11x + 7$, Galois group A_4,
integral basis $\{1, \xi, \xi^2, (1 + \xi^3)/2\}$, $n = 2$, $m(K) = 2$,
$m = 2$, $F(u, v) = u^3 + 16u^2v - 17uv^2 - 576v^3 = \pm 64$
  $F_1(p, q) = p^4 + 3p^3q - 13p^2q^2 - 42pq^3 - 20q^4 = \pm 1, \pm 4$
Minimal index: $\mu(K) = 2$
Solutions: $(1,0,0),(2,2,-1),(7,1,-1),(406,15,-50)$

6.2 Simplest quartic fields

Let $a \in \mathbb{Z}$ with $a \neq 0, \pm 3$. Consider a root ϑ of the polynomial

$$f(x) = x^4 - ax^3 - 6x^2 + ax + 1.$$

The totally real cyclic quartic fields $K = \mathbb{Q}(\vartheta)$ are called *simplest quartic fields*.
    Applying Theorem 6.1.1 G. Lettl and A. Pethő [LP95] described the power
integral bases in the polynomial order $\mathbb{Z}[\vartheta]$ of K.

Theorem 6.2.1 *Up to equivalence all generators of power integral bases of the ring* $\mathbb{Z}[\vartheta]$ *are the following:*

    for arbitrary a $\alpha = \vartheta,\ \alpha = 6\vartheta + a\vartheta^2 - \vartheta^3;$

    for a = 1 additionally $\alpha = 3\vartheta^2 - \vartheta^3,\ \alpha = 25\vartheta + 2\vartheta^2 - 4\vartheta^3;$

    for a = 4 additionally $\alpha = 9\vartheta - 22\vartheta^2 + 4\vartheta^3,\ \alpha = -74\vartheta + 30\vartheta^2 + 9\vartheta^3.$

6.3 An interesting application to mixed dihedral quartic fields

In a recent paper A.C.Kable [Ka99] studied quartic fields K with dihedral Galois group, having power integral basis. These fields have a unique quadratic subfield M. In the following the discriminants of K and M will be denoted by D_K and D_M, respectively.

The main theorem of [Ka99] gives necessary and sufficient conditions for K containing the subfield M to have a power integral basis. The condition is general but does not allow us to settle the problem of existence of power integral bases in K without further analysis. The result has two important consequences (cf. Corollaries 1 and 2 of [Ka99]).

Lemma 6.3.1 *Let K be a dihedral quartic field containing the quadratic subfield M. If K has a power integral basis, then, with a suitable choice of sign, $D_K \pm 4D_M^3$ is a square.*

In the special case of dihedral quartic fields with mixed signature the discriminant is negative and the quadratic subfield is real. Hence

Lemma 6.3.2 *Let K be a mixed dihedral quartic field containing the quadratic subfield M. If K has a power integral basis, then $|D_K| \leq 4D_M^3$. In particular there are only finitely many mixed dihedral quartic fields having a power integral basis and containing a given real quadratic subfield.*

For a given totally real quadratic field M we can enumerate all mixed dihedral quartic fields K satisfying

$$D_M^2 | D_K, \quad D_M^2 \leq |D_K| \leq 4D_M^3 \tag{6.26}$$

(the other properties are implied by $M \subset K$) and then we can select those fields K containing M as subfield and having power integral bases. This way we can determine all mixed dihedral quartic fields K having power integral bases and containing the given real quadratic field M as a subfield. This was performed by I. Gaál and G. Nyul [GNy01]. In Table 11.2.4 all monogenic mixed dihedral extensions of $\mathbb{Q}(\sqrt{2})$, $\mathbb{Q}(\sqrt{3})$ and $\mathbb{Q}(\sqrt{5})$ are listed.

6.4   Totally complex quartic fields

Let again $K = \mathbb{Q}(\xi)$ be a quartic number field and $f(x) = x^4 + a_1 x^3 + a_2 x^2 + a_3 x + a_4 \in \mathbb{Z}[x]$ the minimal polynomial of ξ. If K is a totally complex quartic field, the resolution of the system of equations (6.4) is much simpler. Namely, from the ternary quadratic forms Q_1 and Q_2 we can easily build a linear combination

$$T_\lambda(x, y, z) = Q_1(x, y, z) + \lambda \cdot Q_2(x, y, z)$$

that is positive definite. Then all solutions of

$$T_\lambda(x, y, z) = u + \lambda v$$

can be obtained by direct enumeration. The following theorem was proved by I. Gaál, A. Pethő and M. Pohst [GPP93] (see also I. Gaál [Ga93]).

Theorem 6.4.1 *Let K be a totally complex quartic field. Then the polynomial $R(x) = F(x, 1)$ has three distinct real roots, $\lambda_1 < \lambda_2 < \lambda_3$. The form $T_\lambda(x, y, z) = Q_1(x, y, z) + \lambda \cdot Q_2(x, y, z)$ is positive definite if and only if $\lambda \in (\lambda_1, \lambda_2)$.*

Proof. Denote by ξ_i the roots of $f(x)$, by η_i the roots of $f_0(y) = f(y - a_1/4) = y^4 + b_2 y^2 + b_3 y + b_4$, then $\xi_i = \eta_i - a_1/4$ $(1 \leq i \leq 4)$. Let

$$r(z) = z^3 - 2b_2 z^2 + (b_2^2 - 4b_4)z + b_3^2$$

the cubic resolvent of $f(t)$.

For totally complex quartic fields we have $D(f) > 0$. The discriminant of the cubic resolvent polynomial r of f is equal to $D(f)$, hence $D(r) > 0$. The cubic polynomial $r(z)$ with real coefficients and positive discriminant has three distinct real roots; assume that these are $\mu_1 < \mu_2 < \mu_3$.

We have (cf. R. Kochendörfer [Koc66])

$$
\begin{aligned}
2\eta_1 &= \sqrt{-\mu_1} + \sqrt{-\mu_2} + \sqrt{-\mu_3}, \\
2\eta_2 &= \sqrt{-\mu_1} - \sqrt{-\mu_2} - \sqrt{-\mu_3}, \\
2\eta_3 &= -\sqrt{-\mu_1} + \sqrt{-\mu_2} - \sqrt{-\mu_3}, \\
2\eta_4 &= -\sqrt{-\mu_1} - \sqrt{-\mu_2} + \sqrt{-\mu_3}.
\end{aligned}
$$

Let $c_0 = -a_2 + a_1^2/4$. Set $g(t) = F(-t, 1)/4$, then it is easily seen that

$$r(t - c_0) = -4g(t),$$

hence for the roots δ_i of $g(t)$ we have $\delta_i = \mu_i + c_0$. By $r(0) \geq 0$ we have $r(0) = -\mu_1 \mu_2 \mu_3 \geq 0$ and there are four potential cases for the sign distribution of the μ_i.

I. If we had $\mu_1 < \mu_2 < \mu_3 \leq 0$, then, because of the correspondence between the roots μ_i of the resolvent polynomial and the roots η_i of f_0, all η_i and therefore all ξ_i are real in contradiction to the field being totally complex.

II. For $\mu_1 < \mu_2 \leq 0 < \mu_3$ the inequality $\mu_1\mu_2\mu_3 \leq 0$ can only hold for $\mu_2 = 0$ in which case $\delta_1 < \delta_2 = c_0 < \delta_3$.

III. In case $\mu_1 \leq 0 < \mu_2 < \mu_3$ we have $\delta_1 \leq c_0 < \delta_2 < \delta_3$.

IV. Finally, $0 < \mu_1 < \mu_2 < \mu_3$ contradicts $\mu_1\mu_2\mu_3 \leq 0$.

The leading coefficient of $g(t)$ is negative. Hence, $g(\lambda)$ is positive exactly for

$$\lambda < \delta_1 (\leq c_0) \quad \text{or} \quad (c_0 \leq)\delta_2 < \lambda < \delta_3. \tag{6.27}$$

By $g(t) = F(-t, 1)/4$ the condition $\delta_2 < \lambda < \delta_3$ is equivalent to $\lambda_1 < \lambda < \lambda_2$.

The matrices of the quadratic forms $Q_1(x, y, z)$ and $Q_2(x, y, z)$ are:

$$Q_1 = \begin{pmatrix} 1 & -\dfrac{a_1}{2} & \dfrac{a_1^2}{2} - a_2 \\[2ex] -\dfrac{a_1}{2} & a_2 & \dfrac{a_3 - a_1 a_2}{2} \\[2ex] \dfrac{a_1^2}{2} - a_2 & \dfrac{a_3 - a_1 a_2}{2} & a_4 + a_2^2 - a_1 a_3 \end{pmatrix},$$

$$Q_2 = \begin{pmatrix} 0 & 0 & -\dfrac{1}{2} \\[2ex] 0 & 1 & -\dfrac{a_1}{2} \\[2ex] -\dfrac{1}{2} & -\dfrac{a_1}{2} & a_2 \end{pmatrix}.$$

The principal minors of the matrix of $Q_1 + \lambda Q_2$ are positive if and only if $\lambda > c_0$ and $g(\lambda) > 0$. By (6.27) we conclude that $Q_1 + \lambda Q_2$ is positive definite if and only if $\delta_2 < \lambda < \delta_3$, that is $\lambda_1 < \lambda < \lambda_2$.   □

6.4.1 Parametric families of totally complex quartic fields

The theorem on totally complex quartic fields makes the resolution of index form equations quite easy in such fields. It enables us to consider even infinite parametric families of fields. I. Gaál [Ga93] studied the following families:

$$\begin{aligned}
\text{I} \quad & f_1(x) & = & \; x^4 + k \quad (k > 0,\ k \neq 4k_0^4), \\
\text{II} \quad & f_2(x) & = & \; x^4 + k^2 x^2 - 2kx + 1, \\
\text{III} \quad & f_3(x) & = & \; x^4 + x^3 + kx^2 + \varepsilon x + 1 \quad (k > 0,\ \varepsilon = \pm 1,\ k \neq 2\ if\ \varepsilon = 1), \\
\text{IV} \quad & f_4(x) & = & \; x^4 + 4kx^3 + 4k^2 x^2 + 8x + 4(k^2 + 2k + 3), \\
\text{V} \quad & f_5(x) & = & \; x^4 + kx^2 + 1 \quad (k \geq 3),
\end{aligned}$$

where k is a parameter taking any integer value with the above restrictions (in family I the parameter k is not allowed to admit values of the form $4k_0^4$ where $k_0 \in \mathbb{Z}$).

By elementary means we showed that all these polynomials (for the parameters involved) are irreducible, totally complex, and we calculated their discriminants and Galois groups. Since the integral bases of the fields $K = \mathbb{Q}(\xi)$, generated by a root ξ of these polynomials, are not known, we considered power integral bases in the orders $\mathcal{O} = \mathbb{Z}[\xi]$. We proved the following theorem:

Theorem 6.4.2 *Let* ξ *be a root of one of the polynomials in the families I–V. The element* $\alpha = a + x\xi + y\xi^2 + z\xi^3$ *(a, x, y, z $\in \mathbb{Z}$) generates a power integral basis* $(1, \alpha, \alpha^2, \alpha^3)$ *of the order* $\mathbb{Z}[\xi]$ *of the field* $K = \mathbb{Q}(\xi)$*, if and only if* $a \in \mathbb{Z}$ *and* (x, y, z) *is equal to one of the following triples or their negatives:*

I. $(1, 0, 0), (0, 0, 1)$ *if* $k = 1$, $(1, 0, 0)$ *if* $k \geq 2$,
II. $(1, 0, 0), (k^2, 0, 1), (k^4 + 1, k, k^2)$,
     further, if $k = 2l (l \in \mathbb{Z})$, *then also* $(8l^4 + 1, l, 2l^2)$,
III. $(1, 0, 0), (1, 1, 1), (1, 1, 0), (0, 1, 0), (0, 0, 1), (1, 0, 1)$ *if* $k = 1, \varepsilon = 1$,
     $(1, 0, 0), (3, 1, 1), (2, 0, 1), (1, 1, 0), (0, 1, 0)$ *if* $k = 3, \varepsilon = 1$,
     $(1, 0, 0), (k, 1, 1)$ *for any other allowed pair* k, ε,
IV. $(1, 0, 0)$ *for any* $k \in \mathbb{Z}$,
V. $(1, 0, 0), (k, 0, 1)$ *for any allowed* k.

Proof. The theorem can be proved essentially by using Theorem 6.4.1. As examples we refer briefly to cases IV and V only.
IV. In this case the equation to be solved is

$$F(u, v) = u^3 - 4k^2 u^2 v - 16(k^2 + 3)uv^2 - 64v^3 = \pm 1.$$

Substituting $v_1 = 4v$ we rewrite the above equation as

$$F_1(u, v_1) = u^3 - k^2 u^2 v_1 - (k^2 + 3)uv_1^2 - v_1^3 = \pm 1.$$

This family of cubic Thue equations (corresponding to the "simplest cubic fields" of D. Shanks [Sh74]) was considered by E. Thomas [Tho90] whose results were completed by M. Mignotte [Mi93]. Lemma 5.2.1 implies that the only solution (u, v_1) of this equation, for which v_1 is divisible by 4 is $(\pm 1, 0)$, that yields $(u, v) = (\pm 1, 0)$. We build

$$Q_1 + Q_2$$
$$= \left(x - 2ky + \left(4k^2 - \frac{1}{2}\right)z\right)^2 + (y + (4 - 3k)z)^2 + \left(3k^2 - \frac{17}{4}\right)z^2 = 1,$$

whence only $u = 1$ is possible. For $|k| \geq 2$ the above form is positive definite and z can only be 0. The two candidates we obtain are $(1, 0, 0)$ and $(2k, 1, 0)$ but the last one fails to satisfy (6.4).
  For $k = 0, \pm 1$ we consider $Q_1 + 2Q_2 = 1$, yielding

$$(x - z)^2 + 2(y + 2z)^2 + 3z^2 = 1 \quad \text{for} \quad k = 0,$$
$$(x - 2y + 3z)^2 + 2(y - z)^2 + 5z^2 = 1 \quad \text{for} \quad k = 1,$$
$$(x + 2y + 3z)^2 + 2(y + 5z)^2 + 5z^2 = 1 \quad \text{for} \quad k = -1,$$

all of them having only the trivial solution $(1, 0, 0)$.

V. In this case we have

$$F(u, v) = (u - 2v)(u + 2v)(u - kv) = \pm 1.$$

In can be easily seen that the only solutions are $(u, v) = (\pm 1, 0)$. Further, the form Q_1 is positive definite,

$$Q_1 = (x - kz)^2 + ky^2 + z^2 = 1,$$

which implies $y = 0$ because of $k \geq 3$. If $x - kz = 1, z = 0$, then we get the trivial solution $(1, 0, 0)$; if $x - kz = 0, z = 1$, we obtain the solution $(k, 0, 1)$ which satisfies also the other equation $Q_2 = 0$. $\square$

6.5   Bicyclic biquadratic number fields

Let m, n be distinct square-free integers, $l = (m, n)$, and define m_1, n_1 by $m = lm_1, n = ln_1$. In this case the quartic field $K = \mathbb{Q}(\sqrt{m}, \sqrt{n})$ has Galois group V_4 (the Klein four group). These fields have a very nice special structure.

6.5.1   Integral basis, index form

The integral basis and discriminant of K was described by K.S. Williams [Will70] according to the following five cases. We add also the corresponding index forms:

Case 1. $m \equiv 1 \pmod 4$, $n \equiv 1 \pmod 4$, $m_1 \equiv 1 \pmod 4$, $n_1 \equiv 1 \pmod 4$
integral basis: $\{1, (1 + \sqrt{m})/2, (1 + \sqrt{n})/2, (1 + \sqrt{m} + \sqrt{n} + \sqrt{m_1 n_1})/4\}$
discriminant: $D_K = (lm_1 n_1)^2$

$$
\begin{aligned}
I(x_2, x_3, x_4) = {} & \left(l\left(x_2 + \frac{x_4}{2}\right)^2 - \frac{n_1}{4}x_4^2 \right)\left(l\left(x_3 + \frac{x_4}{2}\right)^2 - \frac{m_1}{4}x_4^2 \right) \\
& \times \left(n_1\left(x_3 + \frac{x_4}{2}\right)^2 - m_1\left(x_2 + \frac{x_4}{2}\right)^2 \right)
\end{aligned}
$$

Case 2. $m \equiv 1 \pmod 4$, $n \equiv 1 \pmod 4$, $m_1 \equiv 3 \pmod 4$, $n_1 \equiv 3 \pmod 4$
integral basis: $\{1, (1 + \sqrt{m})/2, (1 + \sqrt{n})/2, (1 - \sqrt{m} + \sqrt{n} + \sqrt{m_1 n_1})/4\}$
discriminant: $D_K = (lm_1 n_1)^2$

$$
\begin{aligned}
I(x_2, x_3, x_4) = {} & \left(l\left(x_2 - \frac{x_4}{2}\right)^2 - \frac{n_1}{4}x_4^2 \right)\left(l\left(x_3 + \frac{x_4}{2}\right)^2 - \frac{m_1}{4}x_4^2 \right) \\
& \times \left(n_1\left(x_3 + \frac{x_4}{2}\right)^2 - m_1\left(x_2 - \frac{x_4}{2}\right)^2 \right)
\end{aligned}
$$

Case 3. $m \equiv 1 \pmod 4$, $n \equiv 2 \pmod 4$
integral basis: $\{1, (1 + \sqrt{m})/2, \sqrt{n}, (\sqrt{n} + \sqrt{m_1 n_1})/2\}$

discriminant: $D_K = (4lm_1n_1)^2$

$$I(x_2, x_3, x_4) = \left(lx_2^2 - n_1x_4^2\right)\left(l\left(x_3 + \frac{x_4}{2}\right)^2 - \frac{m_1}{4}x_4^2\right)$$
$$\times \left(4n_1\left(x_3 + \frac{x_4}{2}\right)^2 - m_1x_2^2\right)$$

Case 4. $m \equiv 2 \pmod 4$, $n \equiv 3 \pmod 4$
integral basis: $\{1, \sqrt{m}, \sqrt{n}, (\sqrt{m} + \sqrt{m_1n_1})/2\}$
discriminant: $D_K = (8lm_1n_1)^2$

$$I(x_2, x_3, x_4) = \left(\frac{1}{2}(2x_2 + x_3)^2 - \frac{n_1}{2}x_4^2\right)\left(2lx_3^2 - \frac{m_1}{2}x_4^2\right)$$
$$\times \left(2n_1x_3^2 - \frac{m_1}{2}(2x_2 + x_4)^2\right)$$

Case 5. $m \equiv 3 \pmod 4$, $n \equiv 3 \pmod 4$
integral basis: $\{1, \sqrt{m}, (\sqrt{m} + \sqrt{n})/2, (1 + \sqrt{m_1n_1})/2\}$
discriminant: $D_K = (4lm_1n_1)^2$

$$I(x_2, x_3, x_4) = \left(l(2x_2 + x_3)^2 - n_1x_4^2\right)\left(lx_3^2 - m_1x_4^2\right)$$
$$\times \left(\frac{n_1}{4}x_3^2 - m_1\left(x_2 + \frac{x_3}{2}\right)^2\right).$$

Note that for integer x_2, x_3, x_4 all factors attain integer values.

We are going to find the minimal $0 < m \in \mathbb{Z}$ for which the equation

$$I(x_2, x_3, x_4) = \pm m \quad \text{in} \quad x_2, x_3, x_4 \in \mathbb{Z} \tag{6.28}$$

is solvable and all solutions of this equation.

Denote by $F_i = F_i(x_2, x_3, x_4)$ the absolute value of the i-th factor of the index form. We must have $F_1F_2F_3 = m$. It is easily seen by direct calculations (see [GPP95]) that

Lemma 6.5.1 *The following relations hold:*

$$\begin{aligned}
\text{Cases } 1, 2, 4: \quad & \pm F_1m_1 \pm F_2n_1 = \pm F_3l, \\
\text{Case } 3: \quad & \pm F_1m_1 \pm 4F_2n_1 = \pm F_3l, \\
\text{Case } 5: \quad & \pm F_1m_1 \pm F_2n_1 = \pm 4F_3l.
\end{aligned}$$

6.5.2 The totally real case

First we consider totally real bicyclic biquadratic fields. Necessary and sufficient conditions for the existence of power integral bases in such fields were given by

M.N. Gras and F. Tanoe [GT95]. An algorithm for finding all generators of power integral bases was worked out by I. Gaál, A. Pethő and M. Pohst [GPP95]. In the following we detail this procedure.

By Lemma 6.5.1 among the three possible equations only two are independent. In order to be able to deal with all cases in a unique way, introduce integer parameters a, b, c, d, f, g and new variables x, y, z according to the following table:

Case	a	b	c	d	f	g	x	y	z
1	n_1	l	$\pm 4F_1$	m_1	l	$\pm 4F_2$	x_4	$2x_2 + x_4$	$2x_3 + x_4$
2	n_1	l	$\pm 4F_1$	m_1	l	$\pm 4F_2$	x_4	$2x_2 - x_4$	$2x_3 + x_4$
3	n_1	l	$\pm F_1$	m_1	l	$\pm 4F_2$	x_4	x_2	$2x_3 + x_4$
4	n_1	l	$\pm 2F_1$	$m_1/2$	$2l$	$\pm F_2$	x_4	$2x_2 + x_4$	x_3
5	n_1	l	$\pm F_1$	m_1	l	$\pm F_2$	x_4	$2x_2 + x_3$	x_3

Note that m_1 is even in Case 4. Then the system of equations consisting of the two independent equations can be written in the unique form

$$
\begin{aligned}
ax^2 - by^2 &= c, \\
dx^2 - fz^2 &= g \quad \text{in } x, y, z \in \mathbb{Z}.
\end{aligned}
$$

This is a system of simultaneous Pellian equations. A. Baker and H. Davenport [BD69] and R.G.E. Pinch [Pi88] solved formerly such systems of equations; a refined method was given in [GPP95] which we describe here.

Multiply the first equation by a, the second by d and observe that in each Case 1–5 we have $ab = n$ and $df = m$. We obtain

$$
\begin{aligned}
(ax)^2 - ny^2 &= ac, \\
(dx)^2 - mz^2 &= dg \quad \text{in } x, y, z \in \mathbb{Z}.
\end{aligned}
$$

Let $N = \mathbb{Q}(\sqrt{n})$, $M = \mathbb{Q}(\sqrt{m})$ and denote by ε and η their fundamental unit with the property $\varepsilon > 1, \eta > 1$. Denote the conjugate of any $\gamma \in N$ ($\delta \in M$) by $\bar{\gamma}$ (resp. $\hat{\delta}$). Determine all non-associated integral elements $\alpha \in N$, $\beta \in M$ with $N_{N/\mathbb{Q}}(\alpha) = ac$, $N_{M/\mathbb{Q}}(\beta) = dg$. The following algorithm must be performed for all elements of a complete set of non-associated elements α, β of the given norm. (This set can be constructed by using KASH [DF97]). We normalize these elements to be positive. Then

$$
\begin{aligned}
ax + \sqrt{n}\,y &= \alpha \varepsilon^k, \\
dx + \sqrt{m}\,z &= \beta \eta^l, \quad\quad (6.29)
\end{aligned}
$$

with some $k, l \in \mathbb{Z}$. Since all terms in (6.29) are positive, we additionally have $k, l \geq 0$. Expressing x from both equations in (6.29) we get

$$
x = \frac{\alpha \varepsilon^k + \bar{\alpha}\bar{\varepsilon}^k}{2a} = \frac{\beta \eta^l + \hat{\beta}\hat{\eta}^l}{2d}, \quad\quad (6.30)
$$

whence

$$
d\alpha\varepsilon^k + d\bar{\alpha}\bar{\varepsilon}^k = a\beta\eta^l + a\hat{\beta}\hat{\eta}^l.
$$

If

$$l \geq \max\left\{1, \frac{\log(2a|\hat{\beta}|)}{\log \eta}\right\} = l_0 \text{ and } k \geq \max\left\{1, \frac{\log(2d|\bar{\alpha}|)}{\log \varepsilon}\right\} = k_0,$$

we have

$$\eta^l \geq 2a|\hat{\beta}|, \quad \varepsilon^k \geq 2d|\bar{\alpha}|,$$

which, in view of $|N_{N/\mathbb{Q}}(\varepsilon)| = 1$, $|N_{M/\mathbb{Q}}(\eta)| = 1$ implies

$$|d\alpha\varepsilon^k - a\beta\eta^l| = |a\hat{\beta}\hat{\eta}^l - d\bar{\alpha}\bar{\varepsilon}^k| \leq 1.$$

Assume $k \geq l$ (the case $k < l$ can be dealt with similarly). The above estimate yields

$$\left| \frac{a\beta\eta^l}{d\alpha\varepsilon^k} - 1 \right| \leq \frac{1}{d\alpha}\varepsilon^{-k} \leq \frac{1}{\varepsilon} \leq \frac{\sqrt{5}-1}{2} < 0.795.$$

Using the inequality $|\log x| < 2|x - 1|$ holding for $|x - 1| < 0.795$ we obtain

$$\left| l\log \eta - k\log \varepsilon + \log \frac{a\beta}{d\alpha} \right| \leq \frac{2}{d\alpha}\varepsilon^{-k}. \tag{6.31}$$

In case $k < l$ we get

$$\left| k\log \varepsilon - l\log \eta + \log \frac{d\alpha}{a\beta} \right| \leq \frac{2}{a\beta}\eta^{-l}. \tag{6.32}$$

For $0 \leq k \leq k_0, 0 \leq l \leq l_0$ we test whether the pair (k, l) yields a value for x in (6.30). Then we consider the values with $k \geq k_0$ or $l \geq l_0$. Combining (6.31) and (6.32) we obtain that in either case

$$\left| k\log \varepsilon - l\log \eta + \log \frac{d\alpha}{a\beta} \right| \leq c \cdot d^{-H}, \tag{6.33}$$

where

$$c = \frac{2}{\min(d\alpha, a\beta)}, \quad d = \min(\varepsilon, \eta), \quad H = \max(k, l).$$

Applying Baker's method (cf. Section 2.1) from this inequality we obtain an upper bound for H of magnitude 10^{20}. Then, using the Baker–Davenport lemma (cf. Lemma 2.2.1 in Section 2.2.1) this upper bound can be reduced. In our examples the reduced bound was between 5 and 25. The values of k, l under the reduced bound can be tested directly.

   Using this efficient algorithm we determined the minimal indices and all elements with minimal index in all 196 totally real bicyclic biquadratic fields with discriminant $< 10^6$. The data are contained in Table 11.2.5.

6.5.3   The totally complex case

If the field $K = \mathbb{Q}(\sqrt{m}, \sqrt{n})$ is totally complex (this is the only remaining case, since there are no mixed fields of this type), then a factor of the index form is a definite quadratic form which makes the resolution of the index form equation easy in all special cases. M.N. Gras and F. Tanoe [GT95] characterized those fields of this type that are monogenic. Recently G. Nyul [Ny01] succeeded in describing all generators of power integral bases in the monogenic fields of this type. Surprisingly it turns out that the coefficients of the generators of power integral bases with respect to the integral bases belong to a set of a few constant vectors.

To formulate these results we need a refined classification of these fields. Namely, it is easily seen that all such biquadratic number fields can be given in the form $\mathbb{Q}(\sqrt{m}, \sqrt{n})$ so that it belongs to one of the following cases:

Case 1 :	$m > 0,$	$n < 0,$	$m \equiv 1$	(mod 4),	$n \equiv 1$	(mod 4)
			$m_1 \equiv 1$	(mod 4),	$n_1 \equiv 1$	(mod 4)
Case 2 :	$m > 0,$	$n < 0,$	$m \equiv 1$	(mod 4),	$n \equiv 1$	(mod 4)
			$m_1 \equiv 3$	(mod 4),	$n_1 \equiv 3$	(mod 4)
Case 3/A :	$m > 0,$	$n < 0,$	$m \equiv 1$	(mod 4),	$n \equiv 2$	(mod 4)
Case 3/B :	$m < 0,$	$n > 0,$	$m \equiv 1$	(mod 4),	$n \equiv 2$	(mod 4)
Case 4/A :	$m > 0,$	$n < 0,$	$m \equiv 2$	(mod 4),	$n \equiv 3$	(mod 4)
Case 4/B :	$m < 0,$	$n > 0,$	$m \equiv 2$	(mod 4),	$n \equiv 3$	(mod 4)
Case 5/A :	$m > 0,$	$n < 0,$	$m \equiv 3$	(mod 4),	$n \equiv 3$	(mod 4)
Case 5/B :	$m < 0,$	$n < 0,$	$m \equiv 3$	(mod 4),	$n \equiv 3$	(mod 4).

Here merely subcases of Cases 3,4,5 were created. In all cases the integral bases and the corresponding index forms are the same as listed at the beginning of the Section.

Theorem 6.5.1 Let $K = \mathbb{Q}(\sqrt{m}, \sqrt{n})$ be a totally complex biquadratic number field represented in one of the forms listed above. In cases 1., 2. and 3/A. there are no power integral bases. In the other cases the necessary and sufficient condition of the existence of power integral bases in K is:

> Case 3/B: $m_1 = -1, l - 4n_1 = -1$ (and by the assumption $n_1 > 0$)
> Case 4/A: $m_1 = 2, n_1 = -1, l = 1,$ so $m = 2$ and $n = -1$
> Case 4/B: $m_1 = -2, l - n_1 = \pm 2$ (and by the assumption $n_1 > 0$)
> Case 5/A: $n_1 = -1, 4l - m_1 = 1$ (and by the assumption $m_1 > 0$)
> Case 5/B: $l = 1, n_1 - m_1 = \pm 4$ (and by the assumption $m_1, n_1 < 0$).

The solutions of the index form equation corresponding to the above integral basis are

Case 3/B		$(x_2, x_3, x_4) =$	$(1, 1, -2), (1, -1, 2)$
Case 4/A		$(x_2, x_3, x_4) =$	$(0, 0, 1), (1, 0, -1)$
Case 4/B		$(x_2, x_3, x_4) =$	$(0, 0, 1), (1, 0, -1)$
Case 5/A	$m = 3, n = -1$	$(x_2, x_3, x_4) =$	$(1, -2, 1), (1, -2, -1)$
			$(0, 1, 0), (1, -1, 0)$
Case 5/A	other fields	$(x_2, x_3, x_4) =$	$(1, -2, 1), (1, -2, -1)$
Case 5/B		$(x_2, x_3, x_4) =$	$(0, 1, 0), (1, -1, 0)$.

Totally complex biquadratic fields and their generators of power integral bases are listed in Table 11.2.6.

6.5.4   The field index of bicyclic biquadratic number fields

For bicyclic biquadratic fields we succeeded in giving a complete characterization of field indices. (Recall that the field index is the greatest common divisor of the indices of the primitive integral elements of the field.)

Hensel's result ([He08], p.280) implies, that the prime factors of the field index are smaller than the degree of the field. Hence for quartic fields K we have $m(K) = 2^\alpha 3^\beta$. Using the explicit formula of the index forms listed in Section 6.5.1 it is easy to give (cf. I. Gaál, A. Pethő and M. Pohst [GPP91c]) a new proof of T. Nakahara's theorem [Nak83]:

Theorem 6.5.2 *The field index of any bicyclic biquadratic number field is of the form* $m(K) = 2^\alpha 3^\beta$ *with* $\alpha \leq 2, \beta \leq 1$.

Proof. We consider the index forms listed in Section 6.5.1 modulo $8 \cdot 9$. For this purpose we let n_1, m_1, l run through all residue classes modulo $32 \cdot 9$ (because in the index forms these parameters are divided by 2 or 4 in some cases). We exclude the triples which do not fit into one of the Cases 1–5 or for which 4 or 9 divides m or n (they must be square free) or 2 or 3 divides both m_1 and n_1 (they must be coprime). For the remaining triples we let x_2, x_3, x_4 run through all residue classes modulo $8 \cdot 9$ independently, and we calculate the greatest common divisors of the values of $I(x_2, x_3, x_4)$ modulo $8 \cdot 9$. This gcd is never divisible by 8 or 9.   □.

Hence the possible values of the field indices are 1, 2, 3, 4, 6, 12. Surprisingly there are fields with field index 12 neither among the 196 totally real bicyclic biquadratic fields with $D_K \leq 10^6$, cf. Table 11.2.5 nor among the 42 totally complex bicyclic biquadratic fields with $D_K \leq 10^4$, cf. Table 11.2.6.

It is now interesting to ask which values of the field index really occur [GPP91c]. For this sake we let m_1, n_1, l run through all possible values modulo 48 (excluding those which do not fit into Cases 1–5 or do not satisfy the conditions mentioned above). For these triples we let x_2, x_3, x_4 run through the residue classes modulo 12 and calculate the gcd of the values of the index form modulo 12. We sorted the possible values according to the Cases 1–5. The result is displayed in the following table.

$m(K)$	Case 1	Case 2	Case 3	Case 4	Case 5
1			$48k + 1, 2, 5$ $5, 2, 1$ 1600 $-1, 2, 3$ 576	$2, 48k + 1, 3$ $2, 3, 1$ 2304 $2, -1, 1$ 256	$48k + 1, 5, 3$ $5, 1, 3$ 3600 $3, -1, 1$ 144
2	$48k + 1, 5, 13$ $5, 13, 1$ 4225 $5, -3, 1$ 225	$48k + 7, 3, 11$ $7, 3, 11$ 53361 $-1, 3, 7$ 441			
3			$48k + 5, 2, 17$ $13, 10, 1$ 270400 $-1, 2, 11$ 7744	$2, 48k + 5, 11$ $10, 7, 1$ 313600 $2, -1, 5$ 6400	$48k + 1, 13, 7$ $13, 1, 7$ 132496 $7, -5, 1$ 19600
4	$48k + 5, 37, 13$ $17, 33, 1$ 314721 $-7, 17, 1$ 14161	$48k + 11, 3, 19$ $11, 19, 3$ 393129 $-1, 15, 7$ 11025			
6	$48k + 1, 13, 37$ $13, 37, 1$ 231361 $-11, 13, 1$ 20449	$48k + 7, 19, 31$ $19, 31, 7$ 16999129 $-1, 23, 11$ 64009			
12	$48k + 5, 5, 29$ $73, 97, 1$ 50140561 $-23, 73, 1$ 2819041	$48k + 7, 7, 31$ $31, 55, 7$ 142444225 $-1, 47, 23$ 1168561			

In the first row we listed the Cases 1–5, in the first column the field indices. No entry means that in the given case that field index does not occur. Otherwise the entries consist of five lines. In the first line, values m_1, n_1, l are given that represent an infinite family of fields belonging to the given case and having the given index: one of the parameters is an arithmetical progression; taking the prime values from the progression we obtain infinitely many fields with the property displayed. The second line gives the parameters m_1, n_1, l of the totally real bicyclic biquadratic field with smallest discriminant having the given property. The third line gives the corresponding discriminant. Finally, the fourth and fifth lines give the same data for the totally complex bicyclic biquadratic field with smallest discriminant having the given property.

From the table we obtain [GPP91c]:

Theorem 6.5.3 *1. The field index of any bicyclic biquadratic number field is odd if and only if its discriminant is even.*
2. If $K = \mathbb{Q}(\sqrt{m}, \sqrt{n})$ is a bicyclic biquadratic number field represented in one of the forms 1–5, then if both $m \equiv 1 \pmod 4$ and $n \equiv 1 \pmod 4$, then $m(K)$ can be 2, 4, 6, 12, otherwise $m(K)$ can only be 1 or 3. All these values occur infinitely often.

From the list of all triples m_1, n_1, l modulo 48 we select those which correspond to a fixed Case 1–5 and fixed field index. Then we observe some divisibility properties that can be checked by testing all triples in the list. This yields (cf. [GPP91c])

Theorem 6.5.4 *Let $K = \mathbb{Q}(\sqrt{m}, \sqrt{n})$ be a bicyclic biquadratic number field represented with distinct square free m, n according to one of the Cases 1–5. Set $d_1 = m_1 - l, d_2 = n_1 - l, d_3 = m_1 - n_1$. We have*

$m(K) = 1$ *if and only if neither 3 nor 4 divides all three differences d_1, d_2, d_3,*

$m(K) = 2$ *if and only if 4 divides all three differences but neither 3 nor 8 does,*

$m(K) = 3$ *if and only if 3 divides all three differences but 4 does not,*

$m(K) = 4$ *if and only if 8 divides all three differences but 3 does not,*

$m(K) = 6$ *if and only if 3 and 4 divide all three differences but 8 does not,*

$m(K) = 12$ *if and only if both 3 and 8 divide all three differences.*

Note that non-isomorphic fields of the same discriminant do not necessarily have the same field index. This theorem was used in Tables 11.2.5, 11.2.6 to calculate the field indices.

The above theorem gives an answer for bicyclic biquadratic fields to Problem 22 of W. Narkiewicz [Nark74] who asks for an explicit formula for the highest power of a given prime dividing $m(K)$.

7
Quintic Fields

We had to invest the best known reduction and enumeration algorithms, many new ideas and our fastest PC-s to be able to solve index form equations in quintic fields. In the interesting case, for totally real quintic fields (with Galois group A_5 or S_5) this computation takes several hours, contrary to the cubic and quartic cases, where to solve the index form equation was a matter of seconds or at most some minutes. The general method is described in Section 7.1. Having read the relatively complicated formulas of this procedure, in Section 7.2 the reader is rewarded with an interesting family of totally real cyclic quintic fields introduced by E.Lehmer.

7.1 Algorithm for arbitrary quintic fields

In this chapter we present a general algorithm for solving index form equations in arbitrary quintic fields. The method is based on solving the corresponding unit equations, cf. Sections 4.1, 4.2. The crucial point in the method is the application of the enumeration method of K. Wildanger, cf. Section 2.3. The algorithm is described by I. Gaál and K. Győry [GGy99].

The possible Galois groups of quintic fields are C_5 (the cyclic group), D_5 (the dihedral group of order 10), M_{20} (the metacyclic group of degree 5), A_5 and S_5 (cf. H. Cohen [Co93]). By Theorem 1.2.1 of M.N. Gras the index form equation (1.2) has no solution for quintic K having cyclic Galois group, except for the case when K is the maximal real subfield of the 11th cyclotomic field. The orders of the groups C_5 and D_5 do not exceed 10, hence in these cases Wildanger's algorithm can be applied without difficulties to solve directly the unit equations corresponding to the index form equation.

We are going to consider in detail the most difficult cases, when the quintic field K is totally real and has Galois group M_{20}, A_5 or S_5. Obviously, in these cases we have the largest number of unknown exponents in the unit equations.

7.1.1 Preliminaries

Let ξ be an integral generator of K with conjugates $\xi^{(1)} = \xi, \xi^{(2)}, \ldots, \xi^{(5)}$ over $\mathbb{Q}$. We set $K^{(i)} = \mathbb{Q}(\xi^{(i)})$. We denote by $\mathbb{Z}_K$ and D_K the ring of integers and discriminant of K, respectively.

Let α be an arbitrary primitive integral element of $K = \mathbb{Q}(\xi)$. We can write α as

$$\alpha = \frac{x_1 + x_2 \xi + \cdots + x_5 \xi^4}{d} \tag{7.1}$$

with $x_1, \ldots, x_5 \in \mathbb{Z}$ and a common denominator $d \in \mathbb{Z}$. Let $\underline{X} = (X_2, \ldots, X_5)$. Consider the linear forms

$$L_{ij}(\underline{X}) = \left(\xi^{(i)} - \xi^{(j)} \right) X_2 + \cdots + \left(\left(\xi^{(i)} \right)^4 - \left(\xi^{(j)} \right)^4 \right) X_5$$

for distinct i, j with $1 \leq i, j \leq 5$. The element α generates a power integral basis in K if and only if $\underline{x} = (x_2, \ldots, x_5) \in \mathbb{Z}^4$ satisfies

$$\prod_{1 \leq i < j \leq 5} L_{ij}(\underline{x}) = \pm d^{10} \sqrt{|D_{K/\mathbb{Q}}|}. \tag{7.2}$$

We apply here the ideas of Section 4.2. Consider the subfield $L_{i,j} = \mathbb{Q}(\xi^{(i)} + \xi^{(j)}, \xi^{(i)}\xi^{(j)})$ of $K^{(i)} K^{(j)}$. The groups M_{20}, A_5 and S_5 are doubly transitive. The field $K^{(i)} K^{(j)}$ is of degree $5 \cdot 4 = 20$ over $\mathbb{Q}$. The elements of $L_{i,j}$ remain fixed under the action $(i, j) \to (j, i)$ of the Galois group, thus $L_{i,j}$ is a proper subfield of $K^{(i)} K^{(j)}$. Since $\mathbb{Q}(\xi^{(i)}, \xi^{(j)})$ is a quadratic extension of $L_{i,j}$, in our case $L_{i,j}$ is of degree 10 over $\mathbb{Q}$. (Note that in our examples we had $\mathbb{Q}(\xi^{(i)} + \xi^{(j)}) = \mathbb{Q}(\xi^{(i)}\xi^{(j)})$.) Hence in the totally real cases we consider the unit rank of $L_{i,j}$ is 9.

Denote by $\lambda^{(i,j)}$ the conjugate of any $\lambda = \lambda^{(1,2)} \in L_{1,2}$ corresponding to $\xi^{(i)} + \xi^{(j)}, \xi^{(i)}\xi^{(j)}$ ($1 \leq i < j \leq 5$) and for simplicity we let $\lambda^{(j,i)} = -\lambda^{(i,j)}$. It follows from (7.1) that

$$\delta = \frac{d(\alpha^{(1)} - \alpha^{(2)})}{\xi^{(1)} - \xi^{(2)}}$$

is an integer in the field $L_{1,2}$. In view of (7.1), equation (7.2) can be written in the form

$$\prod_{1 \leq i < j \leq 5} \delta^{(i,j)} = \pm d^{10} \frac{\sqrt{|D_{K/\mathbb{Q}}|}}{I(\xi)}.$$

This is just a norm equation in $L_{1,2}$ over $\mathbb{Q}$. Hence there exist an integer γ of norm $\pm d^{10}\sqrt{|D_{K/\mathbb{Q}}|}/I(\xi)$ and a unit η in $L_{1,2}$ such that

$$\delta^{(i,j)} = \gamma^{(i,j)}\eta^{(i,j)} \tag{7.3}$$

for any i, j with $1 \le i < j \le 5$. Note that the following computations must be performed for a complete set of non-associated elements γ of the given norm.

7.1.2 Baker's method, reduction

For any distinct i, j, k, by Siegel's identity we have

$$L_{ij}(\underline{X}) + L_{jk}(\underline{X}) + L_{ki}(\underline{X}) = 0. \tag{7.4}$$

Put

$$\tau^{(ijk)} = \frac{\gamma^{(i,j)}\left(\xi^{(i)} - \xi^{(j)}\right)}{\gamma^{(i,k)}\left(\xi^{(i)} - \xi^{(k)}\right)}, \quad \rho^{(ijk)} = \frac{\eta^{(i,j)}}{\eta^{(i,k)}}.$$

From (7.1), (7.3) and (7.4) we obtain

$$\tau^{(ijk)}\rho^{(ijk)} + \tau^{(kji)}\rho^{(kji)} = 1. \tag{7.5}$$

Denote by $\varepsilon_1, \ldots, \varepsilon_9$ a set of fundamental units in $L_{1,2}$. Then there are integer exponents $a_1, \ldots, a_9$ such that

$$\eta^{(i,j)} = \pm\left(\varepsilon_1^{(i,j)}\right)^{a_1} \cdots \left(\varepsilon_9^{(i,j)}\right)^{a_9}.$$

For $1 \le h \le 9$ let

$$v_h^{(ijk)} = \frac{\varepsilon_h^{(i,j)}}{\varepsilon_h^{(i,k)}},$$

then

$$\rho^{(ijk)} = \left(v_1^{(ijk)}\right)^{a_1} \cdots \left(v_9^{(ijk)}\right)^{a_9}.$$

We have

$$a_1 \log\left|v_1^{(ijk)}\right| + \cdots + a_9 \log\left|v_9^{(ijk)}\right| = \log\left|\rho^{(ijk)}\right|. \tag{7.6}$$

Since $\varepsilon_1, \ldots, \varepsilon_9$ are fundamental units of $L_{1,2}$, hence all 9th order minors of the 10 by 9 matrix $(\log|\varepsilon_h^{(i,j)}|)$ (for $1 \le i < j \le 5, 1 \le h \le 9$) are non-zero and the sum of the row vectors of this matrix is zero, we obtain that the columns of the system of linear equations (7.6) (taken for all distinct $1 \le i, j, k \le 5$) are linearly independent. Let M be a non-zero minor of the matrix $(\log|v_h^{(ijk)}|)$. By multiplication by M^{-1} we can express $a_1, \ldots, a_9$ and we get

$$A = \max_{1 \le h \le 9} |a_h| \le c_1 \max \left|\log\left|\rho^{(i_0 j_0 k_0)}\right|\right|,$$

where c_1 is the row norm of M^{-1} (the maximum sum of the absolute values of the elements in the rows of M^{-1}) and i_0, j_0, k_0 is the triple for which $\left|\log \left|\rho^{(ijk)}\right|\right|$ attains its maximum.

Note that the nine equations should be selected so that c_1 becomes as small as possible. Now if $|\rho^{(i_0 j_0 k_0)}| < 1$, then $\log |\rho^{(i_0 j_0 k_0)}| \leq -A/c_1$ and if $|\rho^{(i_0 j_0 k_0)}| > 1$, then the same holds for $\rho^{(i_0 k_0 j_0)} = 1/\rho^{(i_0 j_0 k_0)}$. Hence we conclude that $|\rho^{(i_0 j_0 k_0)}|$ is small for a certain triple (i_0, j_0, k_0). For the sake of simplicity we omit the subindices in the following, that is we assume

$$\log |\rho^{(ijk)}| \leq -\frac{A}{c_1}.$$

Set $c_2 = |\tau^{(ijk)}|$. Then, using the inequality $|\log x| \leq 2|x - 1|$, holding for $|x - 1| < 0.795$, we deduce from (7.5) that

$$\left\| \log \left|\tau^{(kji)}\right| + a_1 \log \left|v_1^{(kji)}\right| + \cdots + a_9 \log \left|v_9^{(kji)}\right| \right\|$$
$$= \left\| \log \left|\tau^{(kji)}\rho^{(kij)}\right| \right\| \leq 2 \cdot \left| 1 - \left|\tau^{(kji)}\rho^{(kij)}\right| \right|$$
$$\leq 2 \cdot \left| 1 - \tau^{(kji)}\rho^{(kij)} \right| = 2 \cdot \left| \tau^{(ijk)}\rho^{(ijk)} \right| \leq 2c_2 \exp\left(-\frac{A}{c_1}\right), \quad (7.7)$$

provided that the right-hand side is < 0.795, but in the opposite case we get a much better estimate for A. In our examples the terms in the above linear form in logarithms were linearly independent over $\mathbb{Q}$, and applying Theorem 2.1.1 of A. Baker and G. Wüstholz we obtained a lower estimate

$$\left\| \log \left|\alpha^{(kji)}\right| + a_1 \log \left|v_1^{(kji)}\right| + \cdots + a_9 \log \left|v_9^{(kji)}\right| \right\| > \exp(-C \log A)$$

with a large constant C. Comparing the upper and lower estimates for the above linear form we get an upper bound A_0 for A. This upper bound can be reduced by applying Lemma 2.2.2 to inequality (7.7).

7.1.3 Enumeration

Let $\mathcal{I}$ be the set of all triples $I = (ijk)$ with distinct $1 \leq i, j, k \leq 5$. Let

$$\tau^{(I)} = \tau^{(ijk)}, \quad v_h^{(I)} = v_h^{(ijk)} \quad (1 \leq h \leq 9)$$

and

$$\beta^{(I)} = \tau^{(I)} \left(v_1^{(I)}\right)^{a_1} \cdots \left(v_9^{(I)}\right)^{a_9}.$$

If $I' = (kji)$, then in view of (7.5) we have

$$\beta^{(I)} + \beta^{(I')} = 1. \quad (7.8)$$

The set $\mathcal{I}$ satisfies (2.6). Let $I^* = \{I_1, \ldots, I_t\}$ be a set of tuples I with the following properties:

1. if $(i, j, k) \in I^*$ then either $(k, i, j) \in I^*$ or $(k, j, i) \in I^*$,

2. if $(i, j, k) \in I^*$ then either $(j, k, i) \in I^*$ or $(j, i, k) \in I^*$,

3. the vectors

$$
\underline{e}_h = \begin{pmatrix} \log \left| v_h^{(I_1)} \right| \\ \vdots \\ \log \left| v_h^{(I_t)} \right| \end{pmatrix} \quad \text{for} \quad h = 1, \ldots, 9
$$

are linearly independent.

These conditions are tantamout to the ones in Section 2.3. Since $\varepsilon_1, \ldots, \varepsilon_9$ are multiplicatively independent, the last condition can also be satisfied if we take sufficiently many tuples. Note that choosing a minimal set of tuples satisfying those conditions reduces the amount of necessary computation considerably. We are now ready to apply the method in Section 2.3.

In our examples the number of exponent tuples enumerated were still very large, hence we also used *sieving* to get rid of many candidate tuples. We calculated a prime p, coprime to D_K, such that the defining polynomial $f(x)$ of the generating element ξ splits completely mod p, that is

$$
f(x) \equiv (x - r_1)(x - r_2)(x - r_3)(x - r_4)(x - r_5) \pmod{p},
$$

with rational integers $r_1, \ldots, r_5$. Hence $r_1, \ldots, r_5$ can be indexed so that for a certain prime ideal $\wp$ in $\mathbb{Z}_K$ lying above p and for any i $(1 \le i \le 5)$ we have

$$
\xi_i \equiv r_i \pmod{\wp}.
$$

Then we can calculate integers $t^{(ijk)}, n_h^{(ijk)} (h = 1, \ldots, 9)$ for each triple (ijk) of distinct indices $1 \le i, j, k \le 5$ with

$$
\tau^{(ijk)} \equiv t^{(ijk)} \pmod{\wp}
$$

and

$$
v_h^{(ijk)} \equiv n_h^{(ijk)} \pmod{\wp} \quad (1 \le h \le 9).
$$

Then equation (7.8) implies

$$
t^{(ijk)} \left(n_1^{(ijk)} \right)^{a_1} \ldots \left(n_9^{(ijk)} \right)^{a_9} + t^{(kji)} \left(n_1^{(kji)} \right)^{a_1} \ldots \left(n_9^{(kji)} \right)^{a_9} \equiv 1 \pmod{p},
$$

which is very easy and fast to test even for large exponents. In our computations only very few exponent vectors survived this test, and usually they were solutions of (7.8).

In the case we consider the Galois group G is doubly transitive, hence it is enough to solve a single unit equation, say for $i = 1, j = 2, k = 3$. Indeed, if this equation is already solved in $a_1, \ldots, a_9$, then we consider the system of linear equations

$$L_{1j}(\underline{x}) = \pm \left(\xi^{(1)} - \xi^{(j)}\right) \gamma^{(1,j)} \left(\varepsilon_1^{(1,j)}\right)^{a_1}, \ldots, \left(\varepsilon_9^{(1,j)}\right)^{a_9} \tag{7.9}$$

in $\underline{x} = (x_2, \ldots, x_5)$ for $j = 2, 3, 4, 5$. These linear equations are conjugate to each other over $\mathbb{Q}$. The linear forms $L_{1j}(\underline{X})$, $j = 2, 3, 4, 5$, being linearly independent, (7.9) enables us to determine the unknowns $\underline{x} = (x_2, \ldots, x_5)$ from the exponent vectors $(a_1, \ldots, a_9)$, and hence the index form equation can be completely solved.

7.1.4   Examples

EXAMPLE 1
Consider the totally real quintic field $K = \mathbb{Q}(\xi)$ where ξ is defined by the polynomial

$$f(x) = x^5 - 5x^3 + x^2 + 3x - 1.$$

This field has discriminant $D_K = 24217 = 61 \cdot 397$, Galois group S_5, and

$$\omega_1 = 1, \omega_2 = \xi, \omega_3 = \xi^2, \omega_4 = \xi^3, \omega_5 = \xi^4 \tag{7.10}$$

is an integral basis. The element $\xi^{(1)} + \xi^{(2)}$ is defined by the polynomial

$$g(x) = x^{10} - 15x^8 + x^7 + 66x^6 + x^5 - 96x^4 - 7x^3 + 37x^2 + 12x + 1.$$

Then the field $L_{1,2} = \mathbb{Q}(\xi^{(1)} + \xi^{(2)}, \xi^{(1)}\xi^{(2)})$ is generated by $\vartheta = \xi^{(1)} + \xi^{(2)}$ only. An integral basis of $L_{1,2}$ is

$$\{1, \vartheta, \vartheta^2, \vartheta^3, \vartheta^4, \vartheta^5, \vartheta^6, \vartheta^7, \vartheta^8,$$
$$(9 + 27\vartheta + 43\vartheta^2 + 20\vartheta^3 + 37\vartheta^4 + 5\vartheta^5 + 32\vartheta^6 + 3\vartheta^7 + 26\vartheta^8 + \vartheta^9)/47\}$$

and the discriminant of $L_{1,2}$ is $D_{L_{1,2}} = 61^3 \cdot 397^3$. The coefficients of the fundamental units of $L_{1,2}$ with respect to the above integral basis are

(21,	107,	192,	−5,	−120,	−40,	84,	20,	30,	−60)
(16,	99,	139,	−56,	−113,	−7,	56,	9,	14,	−30)
(10,	4,	65,	197,	85,	−110,	56,	34,	50,	−90)
(21,	35,	196,	346,	94,	−206,	129,	66,	97,	−177)
(0,	−53,	−31,	200,	145,	−90,	14,	24,	35,	−60)
(8,	24,	40,	33,	−1,	−27,	25,	10,	15,	−28)
(15,	13,	118,	248,	78,	−143,	84,	45,	66,	−120)
(0,	1,	0,	0,	0,	0,	0,	0,	0,	0)
(4,	19,	42,	0,	−26,	−8,	17,	4,	6,	−12)

Note that the element $\xi^{(1)} \cdot \xi^{(2)}$ has coefficients

$$(-26, -26, -197, -410, -130, 238, -140, -75, -110, 200)$$

in the above integral basis of $L_{1,2}$.

Baker's method (cf. Section 2.1) gave the bound $A_0 = 10^{82}$ for A. This bound was reduced according to the following table:

Step	A_0	H	new bound
I.	10^{82}	10^{900}	3196
II.	3196	10^{55}	205
III.	205	10^{43}	163
IV.	163	$2 \cdot 10^{40}$	153
V.	153	$2 \cdot 10^{35}$	133

Here H is the constant used in Lemma 2.2.2. In the first reduction step we had to use 1300 digits precision, in the following steps 100 digits were enough. We had to perform the reduction in 30 possible cases for the indices (k, j, i). The CPU time for the first step was about 10 hours. The following steps took only some minutes. The final reduced bound 133 gave $S_0 = 10^{691}$ (cf. (2.10)) to start the final enumeration.

For the final enumeration we used the set of 15 ellipsoids defined by

$$I^* = \{ \; (1,2,3), (2,1,3), (3,1,2), (1,2,4), (2,1,4), (4,1,2),$$
$$(1,2,5), (2,1,5), (5,1,2), (1,3,4), (3,1,4), (4,1,3),$$
$$(3,4,5), (4,5,3), (5,3,4) \; \}.$$

Parallel to the enumeration we used sieving modulo $p = 3329$, which was suitable since

$$f(x) \equiv (x + 1752)(x + 1067)(x + 1695)(x + 379)(x + 1765) \pmod{3329}.$$

In the following table we summarize the final enumeration using the ellipsoid method (cf. Section 2.3). In the table we display S, s, the approximate number of exponent vectors $(a_1, \ldots, a_9)$ enumerated in the 15 ellipsoids, and the number of the exponent vectors that survived the modular test. The last line represents the enumeration of the single ellipsoid (2.17).

Step	S	s	enumerated	survived
I.	10^{691}	10^{50}	0	0
II.	10^{50}	10^{20}	0	0
III.	10^{20}	10^{10}	$15 \cdot 5000$	94
IV.	10^{10}	10^{8}	$15 \cdot 1900$	39
V.	10^{8}	10^{6}	$15 \cdot 30000$	532
VI.	10^{6}	10^{5}	$15 \cdot 30000$	563
VII.	10^{5}	10^{4}	$15 \cdot 72000$	1413
VIII.	10000	2500	$15 \cdot 50000$	946
IX.	2500	500	$15 \cdot 66000$	1300
X.	500	100	$15 \cdot 53000$	1032
XI.	100	0	1792512	2135

Steps I-II were very fast, then II-IV took about one hour, V-X about two hours each. The last step XI was again very time consuming, taking about 8 hours of CPU time. We believe that using a finer splitting of the interval the CPU time can be slightly improved, but at least 8 hours of CPU time is necessary.

From the surviving exponent vectors we calculated the solutions of the index form equation corresponding to the basis (7.10):

$(x_2, x_3, x_4, x_5) =$

$(0, 1, 0, 0)$, $(0, 2, 1, -1)$, $(0, 4, 0, -1)$, $(0, 5, 0, -1)$,

$(1, -5, 0, 1)$, $(1, -4, 0, 1)$, $(1, -1, 0, 0)$, $(1, 0, 0, 0)$,

$(1, 1, -2, -1)$, $(1, 4, 0, -1)$, $(2, -1, -1, 0)$, $(2, 4, -1, -1)$,

$(2, 9, -1, -2)$, $(2, 15, -1, -3)$, $(2, 10, -1, -2)$, $(3, 4, -1, -1)$,

$(3, 5, -1, -1)$, $(3, 9, -1, -2)$, $(3, 10, -1, -2)$, $(3, 14, -1, -3)$,

$(3, 18, -2, -4)$, $(4, -1, -1, 0)$, $(4, 0, -1, 0)$, $(4, 5, -1, -1)$,

$(4, 24, -2, -5)$, $(4, 29, -2, -6)$, $(5, -4, -1, 1)$, $(5, 8, -2, -2)$,

$(5, 33, -2, -7)$, $(7, 5, -2, -1)$, $(7, 9, -2, -2)$, $(7, 14, -2, -3)$,

$(9, 18, -3, -4)$, $(11, -13, -2, 3)$, $(12, 27, -4, -6)$, $(17, 28, -6, -6)$,

$(33, 30, -51, -26)$, $(83, 170, -25, -39)$, $(124, 246, -40, -55)$.

EXAMPLE 2

Consider now the totally real quintic field $K = \mathbb{Q}(\xi)$ where ξ is defined by the polynomial

$$f(x) = x^5 - 6x^3 + x^2 + 4x + 1.$$

This field has discriminant $D_K = 36497$ (this is a prime), Galois group S_5, and

$$\omega_1 = 1, \omega_2 = \xi, \omega_3 = \xi^2, \omega_4 = \xi^3, \omega_5 = \xi^4 \qquad (7.11)$$

is an integral basis. The element $\xi^{(1)} + \xi^{(2)}$ is defined by the polynomial

$$g(x) = x^{10} - 18x^8 + x^7 + 96x^6 - 23x^5 - 169x^4 + 44x^3 + 93x^2 - 21x - 11.$$

An integral basis of the field $L_{1,2}$ generated by $\vartheta = \xi^{(1)} + \xi^{(2)}$ is

$$\{1, \vartheta, \vartheta^2, \vartheta^3, \vartheta^4, \vartheta^5, \vartheta^6, \vartheta^7, \vartheta^8,$$
$$(44074 + 62732\vartheta + 54220\vartheta^2 + 50326\vartheta^3 + 32569\vartheta^4 + 35601\vartheta^5$$
$$+31671\vartheta^6 + 29542\vartheta^7 + 84711\vartheta^8 + \vartheta^9)/79083\}$$

and the discriminant of $L_{1,2}$ is $D_{L_{1,2}} = 36497^3$. The coefficients of the fundamental units of $L_{1,2}$ with respect to the above integral basis are

$$
\begin{array}{rrrrrrrrrr}
(456, & 651, & 564, & 527, & 340, & 367, & 328, & 307, & 88, & -821) \\
(3077, & 4375, & 3797, & 3534, & 2273, & 2480, & 2214, & 2066, & 592, & -5527) \\
(7000, & 9968, & 8645, & 8026, & 5166, & 5648, & 5040, & 4701, & 1347, & -12577) \\
(4354, & 6185, & 5339, & 4980, & 3222, & 3504, & 3124, & 2917, & 836, & -7804) \\
(457, & 651, & 564, & 527, & 340, & 367, & 328, & 307, & 88, & -821) \\
(3559, & 5061, & 4378, & 4077, & 2629, & 2867, & 2558, & 2387, & 684, & -6386) \\
(4171, & 5937, & 5144, & 4773, & 3075, & 3366, & 3002, & 2799, & 802, & -7489) \\
(4642, & 6606, & 5716, & 5308, & 3423, & 3743, & 3338, & 3113, & 892, & -8329) \\
(151, & 212, & 182, & 176, & 115, & 120, & 107, & 101, & 29, & -270).
\end{array}
$$

Note that the element $\xi^{(1)} \cdot \xi^{(2)}$ has coefficients

$$(-4354, -6185, -5339, -4980, -3222, -3504, -3124, -2917, -836, 7804)$$

in the above integral basis of $L_{1,2}$.

Baker's method gave the bound $A_0 = 10^{83}$ for A. This bound was reduced according to the following table:

Step	A_0	H	new bound
I.	10^{83}	10^{900}	4078
II.	4078	10^{55}	263
III.	263	10^{44}	214
IV.	214	10^{42}	204

The reduction took about the same CPU time as in Example 1. The final reduced bound 204 gave $S_0 = 10^{1545}$ (cf. (2.10)) to start the final enumeration.

For the final enumeration we used the set of the same 15 ellipsoids as in Example 1.

Parallel to the enumeration we used sieving modulo $p = 2819$, which was suitable since

$$f(x) \equiv (x + 573)(x + 2401)(x + 926)(x + 2266)(x + 2291) \pmod{2819}.$$

In the following table we summarize the final enumeration using the ellipsoid method. The notation is the same as in Example 1.

Step	S	s	enumerated	survived
I.	10^{1545}	10^{50}	0	0
II.	10^{50}	10^{20}	0	0
III.	10^{20}	10^{15}	0	0
IV.	10^{15}	10^{10}	$15 \cdot 200$	2
V.	10^{10}	10^{8}	$15 \cdot 800$	12
VI.	10^{8}	10^{6}	$15 \cdot 13000$	299
VII.	10^{6}	10^{5}	$15 \cdot 13500$	288
VIII.	10^{5}	10^{4}	$15 \cdot 30000$	634
IX.	10000	2500	$15 \cdot 20000$	445
X.	2500	500	$15 \cdot 28000$	624
XI.	500	100	$15 \cdot 22000$	515
XII.	100	0	711746	992

Here the necessary CPU time was somewhat less than in Example 1, which can be seen by looking at the number of vectors tested.

From the surviving exponent vectors we calculated the solutions of the index form equation corresponding to the basis (7.11):

$(x_2, x_3, x_4, x_5) =$

$(1, -6, 0, 1)$, $(1, 0, 0, 0)$, $(2, -6, 0, 1)$, $(2, -5, 0, 1)$,

$(3, -11, 0, 2)$, $(3, -5, 0, 1)$, $(3, 0, -5, 2)$, $(4, -5, -1, 1)$,

$(4, 0, -3, -1)$, $(4, 5, -1, -1)$, $(6, -6, -1, 1)$, $(6, 15, -2, -3)$,

$(7, -12, -1, 2)$, $(7, -11, -1, 2)$, $(8, -12, -1, 2)$, $(9, -18, -1, 3)$,

$(9, -17, -1, 3)$, $(11, -23, -1, 4)$, $(13, -18, -2, 3)$, $(15, -24, -2, 4)$,

$(16, -23, -2, 4)$, $(19, -41, -2, 7)$, $(31, -46, -4, 8)$, $(53, 62, -14, -13)$,

$(80, -159, -9, 27)$, $(115, -166, -15, 29)$.

7.2   Lehmer's quintics

Let $n \in \mathbb{Z}$ and denote by ϑ_n a root of the polynomial

$$f_n(x) = x^5 + n^2 x^4 - (2n^3 + 6n^2 + 10n + 10)x^3$$
$$+ (n^4 + 5n^3 + 11n^2 + 15n + 5)x^2 + (n^3 + 4n^2 + 10n + 10)x + 1. \quad (7.12)$$

These polynomials were discussed by Emma Lehmer [Le88]. Let $K_n = \mathbb{Q}(\vartheta_n)$, ϑ_n a root of f_n.

The fields K_n were also investigated by R. Schoof and L. Washington [SW88] and H. Darmon [Da91] for prime conductors $m = n^4 + 5n^3 + 15n^2 + 25n + 25$.

Under more general conditions, namely assuming only that $m = n^4 + 5n^3 + 15n^2 + 25n + 25$ is square free, I. Gaál and M. Pohst [GP97] considered this family of quintic fields. In order to be able to consider the solutions of the index form equations, we describe an integral basis and a set of fundamental units of the field $K_n = \mathbb{Q}(\vartheta_n)$ in a parametric form. This makes possible to construct explicitly the index form. The most interesting phenomena of the method below (cf. [GP97]) is that in order to prove the non-existence of the solutions of the index form equations (which in fact follows also from Theorem 1.2.1 of M.N. Gras), we purely use congruence considerations modulo m. We believe that these ideas (avoiding time consuming computations) can be useful also for similar types of decomposable form equations.

Some of the symbolic calculations involved here are not possible to perform by hand. In such cases we used Maple (cf. [CG88]). Short proofs often involve tedious symbolic computations.

7.2.1 Integer basis, unit group

In the sequel we frequently use two integers related to the number fields under consideration:

$$\begin{aligned} m &:= n^4 + 5n^3 + 15n^2 + 25n + 25, \\ d &:= n^3 + 5n^2 + 10n + 7, \end{aligned} \tag{7.13}$$

where m is the conductor of the field K_n and d will turn out to be the index of the equation order of f_n in the maximal order under appropriate premises. It is easily seen by Euclid's algorithm that $(m, d) = 1$ for every $n \in \mathbb{Z}$.

For simplicity we denote by $\vartheta = \vartheta_n$ a root of f_n of (7.12) and set $K = K_n = \mathbb{Q}(\vartheta_n)$. As we mentioned before, this field is totally real, cyclic. The Galois group is generated by the transformation

$$x \longmapsto x' = \frac{(n+2) + nx - x^2}{1 + (n+2)x} \tag{7.14}$$

(cf. [SW88]). The following lemma describes the integral basis of K:

Lemma 7.2.1 *If $p^2 \nmid m$ for any prime $p \neq 5$, then an integral basis of K is given by $\{1, \vartheta, \vartheta^2, \vartheta^3, \omega_5\}$ with*

$$\omega_5 = \frac{1}{d}\left\{(n+2) + (2n^2 + 9n + 9)\vartheta + (2n^2 + 4n - 1)\vartheta^2 + (-3n - 4)\vartheta^3 + \vartheta^4\right\}.$$

The discriminant of K is

$$D_K = m^4. \tag{7.15}$$

Proof. Using (7.14) we have $\vartheta' = \alpha + (n+2)^2\omega_5$ with

$$\alpha = (n+2) + (n^3 + 4n^2 + 5n)\vartheta + (-2n^2 - 6n - 5)\vartheta^2 + (n+2)\vartheta^3.$$

The element ω_5 is an algebraic integer in K, since $(n+2, d) = 1$.

The discriminant of the generating polynomial of ϑ is

$$d^2 m^4 = D(\vartheta) = (I(\vartheta))^2 D_K.$$

We shall show that the index $I(\vartheta) = (\mathbb{Z}_K^+ : \mathbb{Z}^+[\vartheta])$ is equal to d, which in view of $D(1, \vartheta, \vartheta^2, \vartheta^3, \omega_5) = m^4$ implies that $\{1, \vartheta, \vartheta^2, \vartheta^3, \omega_5\}$ is indeed an integral basis. The inclusion

$$\mathbb{Z}_K \supseteq \mathbb{Z}[\vartheta, \omega_5] \supseteq \mathbb{Z}[\vartheta] \quad \text{and} \quad (\mathbb{Z}^+[\vartheta, \omega_5] : \mathbb{Z}^+[\vartheta]) = d$$

shows that d divides $I(\vartheta)$.

In view of $(m, d) = 1$ we must still show that no prime number p dividing m occurs in that index. We discuss the cases $p \neq 5$ and $p = 5$ separately.

I. Let us assume that $p \neq 5$ at first. Then $-n^2/5$ is a five-fold zero of $f_n(x)$ modulo $p\mathbb{Z}[x]$. We get

$$f_n(x) - \left(x + \frac{n^2}{5}\right)^5 = \sum_{i=1}^{4} b_i x^{4-i}$$

with

$$b_1 = -\frac{2}{5}n^4 - 2n^3 - 6n^2 - 10n - 10,$$

$$b_2 = -\frac{2}{25}n^6 + n^4 + 5n^3 + 11n^2 + 15n + 5,$$

$$b_3 = -\frac{1}{125}n^8 + n^3 + 4n^2 + 10n + 10,$$

$$b_4 = -\frac{1}{3125}n^{10} + 1.$$

Let

$$\tilde{m} = \begin{cases} m & \text{for } 5 \nmid m, \\ \dfrac{m}{25} & \text{for } 5 \mid m, \end{cases}$$

then obviously $5^i b_i \equiv 0 \pmod{\tilde{m}}$, hence

$$3125 \left[f_n(x_0) - \left(x_0 + \frac{n^2}{5}\right)^5 \right]_{x_0 = -\frac{n^2}{5}} = mk$$

with $k = 4n^6 + 3n^5 + 65n^4 - 200n^2 - 125n + 125$. Another gcd computation shows that $(\tilde{m}, k) = 1$. Hence, the Dedekind test (Ch. 4.5 (5.55) in M. Pohst and H. Zassenhaus [PZ89]) implies that $R := \mathbb{Z}[1, \vartheta, \vartheta^2, \vartheta^3, \omega_5]$ is p–maximal exactly for $p^2 \nmid m$.

II. Finally, we consider the case $p = 5$. Clearly, $5 \mid n$. Setting $n = 5\tilde{n}$ we obtain $m = 5^2 \tilde{m}$ with $\tilde{m} = 5(5\tilde{n}^4 + 5\tilde{n}^3 + 3\tilde{n}^2 + \tilde{n}) + 1$. An easy calculation shows that $f_n(x) \equiv x^5 - 10x^3 + 5x^2 + 10x + 1 \pmod{25\mathbb{Z}[x]}$ and therefore

$f_n(x) \equiv (x+1)^5 \pmod{5\mathbb{Z}[x]}$. For the Dedekind test we must check whether -1 is a zero of $h_n(x) = (f_n(x) - (x+1)^5)/5$ in $\mathbb{Z}/5\mathbb{Z}$. Because

$$h_n(-1) = 5^{-1}(n^4 + 6n^3 + 14n^2 + 15n + 15) \equiv 3 \pmod{5},$$

we see that R is 5–maximal.

The discriminant can be calculated directly. $\qquad\qquad\qquad\qquad\qquad$ $\square$

The arguments imply also that except for $n = -1, -2$ the order $R := \mathbb{Z}[\vartheta, \omega_5]$ is strictly larger than $\mathbb{Z}[\vartheta]$. The order R is the maximal order of K if and only if there is no prime number $p \neq 5$ whose square divides m.

Now we describe a set of fundamental units of K:

Lemma 7.2.2 *Any four distinct roots of f_n form a set of fundamental units in K.*

Proof. This statement was proved by R. Schoof and L. Washington ([SW88], Theorem 3.5) for the case that m is a prime number. We follow the arguments of their proof and recall only the major steps. By i we denote the index of the subgroup generated by four roots of $f_n(x)$ and -1 in the full unit group of K.

(i) for $|n+1| \geq 20$ we have $i < 11$.

(ii) Since 2, 3, 7 and 9 are not norms for $\mathbb{Z}[\zeta_5]$ over $\mathbb{Z}$ (ζ_5 being a fifth primitive root of unity), we have $i \in \{1, 5\}$ for $|n+1| \geq 20$.

(iii) The possibility $i = 5$ is eliminated by considering a prime $p \neq 5$ dividing m (compare Step 2 of [SW88]). Such a prime exists except for $n = 0$.

(iv) For those n subject to $|n+1| < 20$ for which m is not a prime the unit group is explicitly calculated by KASH (cf. [DF97]). $\qquad\qquad\qquad\qquad$ $\square$

7.2.2 The index form

We are now ready to consider the index form corresponding to the integral basis given in Lemma 7.2.1. Denote by $\gamma^{(i)}$ ($1 \leq i \leq 5$) the conjugates of any $\gamma \in K$ ordered such that (7.14) maps $\gamma^{(i)}$ onto $\gamma^{(j)}$, $j = (i \mod 5) + 1$. For $1 \leq i \leq 5$ we set

$$L^{(i)}(\underline{X}) = \vartheta^{(i)} X_2 + \left(\vartheta^{(i)}\right)^2 X_3 + \left(\vartheta^{(i)}\right)^3 X_4 + \omega_5^{(i)} X_5,$$

and

$$L_{ij}(\underline{X}) = L^{(i)}(\underline{X}) - L^{(j)}(\underline{X}) \quad (1 \leq i < j \leq 5)$$

in the variables $\underline{X} = (X_2, X_3, X_4, X_5)$. The index form corresponding to the integral basis of Lemma 7.2.1 is

$$I(\underline{X}) = \frac{1}{\sqrt{D_K}} \prod_{1 \leq i < j \leq 5} L_{ij}(\underline{X}). \qquad (7.16)$$

This is a homogeneous form of degree 10 with rational integer coefficients. In view of the following lemma it splits into two norm forms of degree 5 each.

Lemma 7.2.3 *We have*

$$I(\underline{X}) = N_{K/\mathbb{Q}}(A(\underline{X}))N_{K/\mathbb{Q}}(B(\underline{X})),$$

where

$$A(\underline{X}) = \frac{L_{12}(\underline{X})}{\vartheta^{(1)} - \vartheta^{(3)}}$$

$$= \frac{\vartheta^{(1)} - \vartheta^{(2)}}{\vartheta^{(1)} - \vartheta^{(3)}} X_2 + \frac{\left(\vartheta^{(1)}\right)^2 - \left(\vartheta^{(2)}\right)^2}{\vartheta^{(1)} - \vartheta^{(3)}} X_3 + \frac{\left(\vartheta^{(1)}\right)^3 - \left(\vartheta^{(2)}\right)^3}{\vartheta^{(1)} - \vartheta^{(3)}} X_4$$

$$+ \frac{\omega_5^{(1)} - \omega_5^{(2)}}{\vartheta^{(1)} - \vartheta^{(3)}} X_5$$

and

$$B(\underline{X}) = \frac{L_{13}(\underline{X})}{\vartheta^{(1)} - \vartheta^{(3)}}$$

$$= X_2 + \left(\vartheta^{(1)} + \vartheta^{(3)}\right) X_3 + \left(\left(\vartheta^{(1)}\right)^2 + \vartheta^{(1)}\vartheta^{(3)} + \left(\vartheta^{(3)}\right)^2\right) X_4$$

$$+ \frac{\omega_5^{(1)} - \omega_5^{(3)}}{\vartheta^{(1)} - \vartheta^{(3)}} X_5$$

and the coefficients of these linear forms are integers of K.

Proof. In view of (7.15) we have

$$I(\underline{X}) = \frac{1}{m^2} N_{K/\mathbb{Q}}(L_{12}(\underline{X})) N_{K/\mathbb{Q}}(L_{13}(\underline{X})).$$

We observe that $N_{K/\mathbb{Q}}(\vartheta^{(1)} - \vartheta^{(3)}) = -m$. By direct calculation it can be checked that the coefficients of both $L_{12}(\underline{X})$ and $L_{13}(\underline{X})$ are divisible by $\vartheta^{(1)} - \vartheta^{(3)}$. □

7.2.3   The index form equation

Using only congruence considerations we show:

Theorem 7.2.1 *Assume that m is square free. The field K admits a power integral basis if and only if $n = -1$ or $n = -2$.*

Proof. Let $\underline{x} = (x_2, x_3, x_4, x_5) \in \mathbb{Z}^4$ and consider

$$\alpha = \vartheta x_2 + \vartheta^2 x_3 + \vartheta^3 x_4 + \omega_5 x_5. \tag{7.17}$$

Obviously, we have

$$\frac{L_{12}(\underline{x})}{\vartheta^{(1)} - \vartheta^{(3)}} + \frac{L_{23}(\underline{x})}{\vartheta^{(1)} - \vartheta^{(3)}} - \frac{L_{13}(\underline{x})}{\vartheta^{(1)} - \vartheta^{(3)}} = 0. \tag{7.18}$$

By Lemma 7.2.3

$$\varepsilon = \frac{L_{12}(x)}{\vartheta^{(1)} - \vartheta^{(3)}} = A(\underline{x}) \text{ and } \eta = \frac{L_{13}(x)}{\vartheta^{(1)} - \vartheta^{(3)}} = B(\underline{x})$$

are units in K. The conjugate of ε under the mapping (7.14) is

$$\varepsilon' = \left(\frac{L_{12}(x)}{\vartheta^{(1)} - \vartheta^{(3)}} \right)' = \frac{L_{23}(x)}{\vartheta^{(2)} - \vartheta^{(4)}},$$

hence we obtain an equation of the form

$$\varepsilon + \left(\frac{\vartheta^{(2)} - \vartheta^{(4)}}{\vartheta^{(1)} - \vartheta^{(3)}} \right) \varepsilon' - \eta = 0, \tag{7.19}$$

where ε' denotes the conjugate of ε under (7.14).

Consider now equation (7.19) modulo m. The assumption m being square free involves $5 \nmid m$ (or equivalently $5 \nmid n$) hence 5 is invertible modulo m. Also, in view of $(m, d) = 1$, d is invertible modulo m. Because of

$$f_n(x) \equiv \left(x + \frac{n^2}{5} \right)^5 \pmod{m}$$

all roots η_i of f_n satisfy

$$\eta_i \equiv -\frac{n^2}{5} \pmod{m} \ (1 \le i \le 5). \tag{7.20}$$

Moreover, we have $\eta_1 \eta_2 \eta_3 \eta_4 \eta_5 = -1$, hence

$$\left(\frac{-n^2}{5} \right)^5 \equiv -1 \pmod{m}. \tag{7.21}$$

Lemma 7.2.2 implies that $\{\eta_1, \eta_2, \eta_3, \eta_4\}$ is a set of fundamental units in K, hence any unit ε can be written as a power product of these elements and possibly -1. Now (7.20) and (7.21) imply that for any unit ε there exists an exponent k $(0 \le k \le 4)$ such that

$$\varepsilon \equiv \pm \left(\frac{-n^2}{5} \right)^k \pmod{m}.$$

The remainder of $(-n^2/5)^k \pmod{m}$ $(0 \le k \le 4)$ is just one of the following values:

$$e_0 = 1, \ e_1 = \frac{-n^2}{5}, \ e_2 = -\frac{n^3}{5} - \frac{3n^2}{5} - n - 1,$$

$$e_3 = \frac{2n^2}{5} + n + 2, \ e_4 = \frac{n^3}{5} + \frac{4n^2}{5} + 2n + 2.$$

By

$$\eta_i \equiv \eta_i' \equiv \frac{-n^2}{5} \pmod{m} \ (1 \le i \le 5).$$

we confirm $\varepsilon \equiv \varepsilon' \pmod{m}$ for any unit ε of K. Let

$$a = \frac{n^3}{5} + \frac{4n^2}{5} + 2n + 2.$$

It is easy to see that

$$\frac{\vartheta^{(2)} - \vartheta^{(4)}}{\vartheta^{(1)} - \vartheta^{(3)}} \equiv a \pmod{m}.$$

Using $\varepsilon \equiv \varepsilon' \pmod{m}$ and (7.19) we obtain

$$\varepsilon(1 + a) - \eta \equiv 0 \pmod{m},$$

whence

$$(\eta)^{-1}\varepsilon(1 + a) - 1 \equiv 0 \pmod{m}.$$

Since $(\eta)^{-1}\varepsilon$ is a unit in K, we obtain

$$\pm e_k(1 + a) - 1 \equiv 0 \pmod{m}, \tag{7.22}$$

with a suitable sign with one of $0 \le k \le 4$.

We calculate the remainders modulo m of the left-hand sides of (7.22) for $0 \le k \le 4$ and for both possible signs. These remainders are cubic polynomials in n. It is easy to see that for $|n| > 250$ they are non–zero and in absolute value less than m. For $|n| \le 250$ we test all these congruences and the only solutions found are $n = -1, -2$. For $n \ne -1, -2$ we get a contradiction, since if α generated a power integral basis in K, then congruence (7.22) would be solvable for a suitable k and a suitable sign. $\qquad\square$

7.2.4   The exceptional case

For $n = -1, -2$ the fields K_n coincide, which is easily checked by KASH [DF97]. To fix our notation we set $n = -1$, $K = K_{-1}$ and $\vartheta = \vartheta_{-1}$. We note that K is the totally real quintic number field of minimum discriminant. We have $m = 11, d = 1$, hence by Lemma 7.2.1 $\{1, \vartheta, \vartheta^2, \vartheta^3, \vartheta^4\}$ is an integral basis of K. Using standard arguments, by combining Baker's method (cf. Section 2.1) with reduction algorithms (cf. Section 2.2.2) and sieving procedures, we solved the unit equation

$$\varepsilon\eta^{-1} + \left(\frac{\vartheta^{(2)} - \vartheta^{(4)}}{\vartheta^{(1)} - \vartheta^{(3)}} \right) \varepsilon'\eta^{-1} = 1$$

implied by (7.19). The solutions allow us to express ε/ε' and hence also ε, which gives (x_2, x_3, x_4, x_5) in view of the representation of (7.17) by taking conjugates and solving the corresponding system of linear equations. We obtained the following solutions:

$(x_2, x_3, x_4, x_5) = (0, 1, 0, 0), (0, 3, 0, -1), (0, 4, 0, -1), (1, -4, 0, 1),$
$(1, -3, 0, 1), (1, -2, -1, 1), (1, -1, -1, 0), (1, 0, 0, 0), (1, 1, 0, 0),$
$(2, -1, -1, 0), (2, 0, -1, 0), (2, 1, -2, -1), (2, 1, -1, 0), (2, 3, -1, -1),$
$(2, 4, -1, -1), (2, 8, -1, -2), (3, -1, -1, 0), (3, 0, -1, 0), (3, 3, -1, -1),$
$(3, 4, -1, -1), (4, -4, -1, 1), (5, -11, -1, 3), (5, 2, -2, -1), (5, 13, -2, -3),$
$(11, 5, -4, -2).$

8
Sextic Fields

An analogue of the general method used for quintic fields, reducing the index form equation directly to unit equations, does not seem to be feasible in sextic fields.

Hence, in the sextic case we can compute generators of power integral bases only if the fields admit some additional property, making the index form equation easier. We have efficient algorithms for sextic fields having quadratic or cubic subfields (see Sections 8.1, 8.2). Investigating the structure of the index form in sextic fields with a quadratic subfield we shall point out the important role of various types of Thue equations (see [Ga96b]).

In Section 8.3 we shall consider sextic fields that are composites of a quadratic and a cubic field. We show some interesting applications of the results of Section 4.4.2 on composite fields. We close the chapter by investigating power integral bases in the infinite parametric family of simplest sextic fields (Section 8.3.3).

8.1   Sextic fields with a quadratic subfield

Let M be a quadratic field with integral basis $\{1, \omega\}$. Let $f(t) = t^3 + \gamma_2 t^2 + \gamma_1 t + \gamma_0 \in \mathbb{Z}_M[t]$ be the minimal polynomial of the sextic integer ϑ, generating the field $K = \mathbb{Q}(\vartheta)$. In the tables of A.M. Bergé, J. Martinet and M. Olivier [BMO90], M. Olivier [Oli89], in about 99% of the cases this ϑ can be chosen so that ϑ has relative index 1 over M, which implies that $\{1, \vartheta, \vartheta^2, \omega, \omega\vartheta, \omega\vartheta^2\}$ is an integral basis of K. To make our formulas simpler in the following we assume that it is indeed an integral basis and any $\alpha \in \mathbb{Z}_K$ can be represented in the form

$$\alpha = x_0 + x_1\vartheta + x_2\vartheta^2 + y_0\omega + y_1\omega\vartheta + y_2\omega\vartheta^2 \tag{8.1}$$

with $x_0, x_1, x_1, y_0, y_1, y_2 \in \mathbb{Z}$. We note however that in the remaining cases we only need to use a common denominator $g \in \mathbb{Z}$ in (8.1) which implies additional constants on the right sides of our formulas, but in principle the same arguments work.

Consider the index form equation

$$I(x_1, x_2, y_0, y_1, y_2) = \pm 1 \quad \text{in} \quad x_1, x_2, y_0, y_1, y_2 \in \mathbb{Z} \tag{8.2}$$

corresponding to the above integral basis.

Let $\vartheta^{(1)}$ and $\vartheta^{(2)}$ be distinct roots of $f(t)$ and put $\rho = -\vartheta^{(1)} - \vartheta^{(2)}$. For a solution $(x_1, x_2, y_0, y_1, y_2)$ of (8.2) set $X = x_1 + \omega y_1$, $Y = x_2 + \omega y_2$. According to the general remarks in Section 4.3 the quadratic subfield implies a factor of the index form (cf. [Ga95], [Ga96a], [GP96]):

Theorem 8.1.1 *If $(x_1, x_2, y_0, y_1, y_2)$ is a solution of (8.2), then $(X, Y) = (x_1 + \omega y_1, x_2 + \omega y_2)$ is a solution of the relative Thue equation*

$$N_{K/M}(X - \rho Y) = \nu \quad \text{in} \quad X, Y \in \mathbb{Z}_M, \tag{8.3}$$

where ν is a unit in M, and

$$F(x_1, x_2, y_0, y_1, y_2) = \pm 1, \tag{8.4}$$

with a homogeneous polynomial F of degree 9 with integer coefficients.

Proof. Denote by $\vartheta = \vartheta^{(1)}, \vartheta^{(2)}, \vartheta^{(3)}$ the conjugates of ϑ over M, denote by $\overline{\omega}$ the conjugate of ω, and arrange the conjugates of any $\gamma \in K$ according to $\omega^{(j)} = \omega$, $\omega^{(j+3)} = \overline{\omega}$ $(1 \le j \le 3)$.

By straightforward computation we obtain

$$\sqrt{|D_K|} = \left| (\omega - \overline{\omega})^3 \right|$$
$$\times \left| N_{M/\mathbb{Q}} \left((\vartheta^{(1)} - \vartheta^{(2)})(\vartheta^{(2)} - \vartheta^{(3)})(\vartheta^{(3)} - \vartheta^{(1)}) \right) \right|. \tag{8.5}$$

Also, we have

$$I(\alpha) = \frac{\left| \prod_{1 \le j < k \le 6} (\alpha^{(j)} - \alpha^{(k)}) \right|}{\sqrt{|D_K|}} = \pm 1. \tag{8.6}$$

Obviously, for $(j, k) = (1,2), (2,3), (1,3)$,

$$\alpha^{(j)} - \alpha^{(k)} = (\vartheta^{(j)} - \vartheta^{(k)}) \left((x_1 + \omega y_1) - (\vartheta^{(j)} + \vartheta^{(k)})(x_2 + \omega y_2) \right),$$

hence, the product of the $|\alpha^{(j)} - \alpha^{(k)}|$ for $(j, k) = (1,2), (2,3), (1,3), (4,5), (5,6), (4,6)$ is equal to

$$N_{M/\mathbb{Q}} \left((\vartheta^{(1)} - \vartheta^{(2)})(\vartheta^{(2)} - \vartheta^{(3)})(\vartheta^{(3)} - \vartheta^{(1)}) \right)$$
$$\times N_{K/\mathbb{Q}}((x_1 + \omega y_1) - \rho(x_2 + \omega y_2)).$$

Dividing by $\sqrt{|D_K|}$ in (8.6), the first factor cancels. The second factor is a primitive polynomial with integer coefficients, hence it divides $I(\alpha)$ in $\mathbb{Z}[x_1, x_2, y_0, y_1, y_2]$. Denote by $F(x_1, x_2, y_0, y_1, y_2)$ the product of the remaining nine factors $|\alpha^{(j)} - \alpha^{(k)}|$ (for the (j, k) not listed above) divided by $(\omega - \overline{\omega})^3$. This $F(x_1, x_2, y_0, y_1, y_2)$ must be equal to the quotient of $I(\alpha)$ by $N_{K/\mathbb{Q}}((x_1 + \omega y_1) - \rho(x_2 + \omega y_2))$, hence it also has rational integer coefficients. Finally, by $I(\alpha) = 1$ we conclude

$$N_{K/\mathbb{Q}}((x_1 + \omega y_1) - \rho(x_2 + \omega y_2)) = \pm 1$$

and (8.4) holds. The above equation implies (8.3).   □

In the following we consider real and imaginary quadratic subfields M separately.

8.1.1 Real quadratic subfield

In case M is a real quadratic subfield, there are infinitely many units ν in M. The relative Thue equation (8.3) can be solved by using the algorithm described in Section 3.3 and determines the variables $X, Y \in \mathbb{Z}_M$ up to unit factors in M. That is we can calculate finitely many $(X, Y) \in \mathbb{Z}_M^2$ such that any solution of (8.3) is of the form $(\mu^n X, \mu^n Y)$, where μ is the fundamental unit in M chosen with the property $|\mu| > 1$. By

$$x_1 + \omega y_1 = \mu^n X, \quad x_2 + \omega y_2 = \mu^n Y$$

we obtain

$$
\begin{aligned}
x_1 &= \frac{\mu^n X \overline{\omega} - (\overline{\mu})^n \overline{X} \omega}{\overline{\omega} - \omega}, \\
y_1 &= \frac{\mu^n X - (\overline{\mu})^n \overline{X}}{\omega - \overline{\omega}}, \\
x_2 &= \frac{\mu^n Y \overline{\omega} - (\overline{\mu})^n \overline{Y} \omega}{\overline{\omega} - \omega}, \\
y_2 &= \frac{\mu^n Y - (\overline{\mu})^n \overline{Y}}{\omega - \overline{\omega}}.
\end{aligned}
\tag{8.7}
$$

Substituting the values of (8.7) into (8.4), we obtain an equation of the form

$$\prod_{k=1}^{9} \left(A_k \mu^n + B_k (\overline{\mu})^n + C_k y_0 \right) = \pm 1, \tag{8.8}$$

with explicitly known algebraic coefficients A_k, B_k, C_k $(1 \le k \le 9)$. Consider this equation for $n \ge 0$. The opposite case of $n < 0$ is similar by interchanging the roles of A_k and B_k. If $n \ge 0$, then in (8.8) the dominating variables are μ^n and y_0, and the value of $(\overline{\mu})^n$ is "small" compared to the dominating variables (we recall that we defined μ with $|\mu| > 1$). The structure of this equation is very similar to that of an inhomogeneous Thue equation considered in Section 3.2. In many respects the situation is much simpler because, except for small $n > 0$ (which values can be tested separately), the value of $|(\bar{\mu})^n|$ can be bounded by a quite small absolute constant. In fact equation (8.8) can be solved by the methods of Section 3.2. Having determined n and y_0 we can also calculate x_1, x_2, y_1, y_2 by (8.7).

8.1.2 Totally real sextic fields with a quadratic and a cubic subfield

If the field K additionally to the quadratic subfield M admits also a cubic subfield L, then the situation becomes simpler. Namely, the polynomial F involved in (8.4) has a cubic factor that makes the resolution of (8.8) much simpler (see [Ga96a]). In this case, by choosing the numeration of the conjugates we also assume having $\vartheta^{(i)}, \vartheta^{(i+3)}$ relative conjugates over the cubic subfield L ($i = 1, 2, 3$).

If $L = \mathbb{Q}(\rho)$, then obviously $\{1, \rho, \rho^2, \omega, \omega\rho, \omega\rho^2\}$ is a (not necessarily integral) basis of K over $\mathbb{Q}$. Using this basis it is easily seen that $\alpha^{(1)} - \alpha^{(4)}$ is in L. Then

$$G_3(x_1, x_2, y_0, y_1, y_2) = (\alpha^{(1)} - \alpha^{(4)})(\alpha^{(2)} - \alpha^{(5)})(\alpha^{(3)} - \alpha^{(6)})$$

is a complete norm, hence it has coefficients in $\mathbb{Z}$. Similarly, the product of the remaining six differences

$$\begin{aligned}
G_6&(x_1, x_2, y_0, y_1, y_2) \\
&= (\alpha^{(1)} - \alpha^{(5)})(\alpha^{(1)} - \alpha^{(6)})(\alpha^{(2)} - \alpha^{(4)}) \\
&\times (\alpha^{(2)} - \alpha^{(6)})(\alpha^{(3)} - \alpha^{(4)})(\alpha^{(3)} - \alpha^{(5)})
\end{aligned}$$

must also have integer coefficients. Note that this later form must also be a norm form. The product of these two forms must be divisible by the remaining factor $(\omega - \bar{\omega})^3$ of $\sqrt{|D_K|}$ in (8.5). We conclude that there must be a cubic element δ_3 of norm d_3 and a sextic element δ_6 of norm d_6 such that δ_3 divides $\alpha^{(1)} - \alpha^{(4)}$, δ_6 divides $\alpha^{(1)} - \alpha^{(5)}$, and $d_3 \cdot d_6 = (\omega - \bar{\omega})^3$. We obtain that

$$F_3(x_1, x_2, y_0, y_1, y_2) = \frac{1}{d_3} G_3(x_1, x_2, y_0, y_1, y_2) = \pm 1,$$

with a homogeneous cubic form F_3 with integer coefficients. Using the substitution (8.7) this gives rise to an equation similar to (8.8) but of degree 3, which is much simpler to solve. A detailed description of the resolution can be found in I. Gaál [Ga96a].

The case considered in this section is especially important because the totally real cyclic sextic fields are of this type. These fields were intensively studied by several authors, see S. Mäki [Ma80], V. Ennola, S. Mäki and R. Turunen [EMT85]. Recently S.I.A. Shah [Sy00] investigated power integral bases in cyclic sextic fields with prime conductor.

In [Ga96a] we computed all power integral bases in the five totally real cyclic sextic fields of smallest discriminants, by using the above described method. The corresponding table is given in Section 11.3.1.

8.1.3   Imaginary quadratic subfield

In case M is an imaginary quadratic subfield of the sextic field K, there are only finitely many units ν in M and the relative Thue equation (8.3) determines completely $X, Y \in \mathbb{Z}_M$. By

$$x_1 + \omega y_1 = X, \quad x_2 + \omega y_2 = Y$$

for all solutions $(X, Y) \in \mathbb{Z}_M^2$ we can determine the corresponding x_1, y_1, x_2, y_2 $\in \mathbb{Z}$. Substituting these values into equation (8.4) we get a polynomial equation of degree 9 in the remaining variable y_0. This way it is easy to solve index form equations corresponding to such fields. This was considered by I. Gaál and M. Pohst [GP96]. The corresponding computational results are given in Section 11.3.2.

8.1.4   Sextic fields with an imaginary quadratic and a real cubic subfield

An especially delicious case is the case of sextic fields K having both an imaginary quadratic subfield M and a real cubic subfield L. The totally complex cyclic sextic fields are special examples for such fields. Power integral bases in number fields of this type were considered by I. Gaál [Ga95]. We show that in this special situation our problem can be reduced to solving cubic Thue inequalities over $\mathbb{Z}$. The method below enables one to determine even all integer elements of K of given index. It can also be applied to infinite parametric families of sextic fields of this type.

In the following ϑ is a totally real cubic algebraic integer and m is a square–free positive integer. Let us consider the field $K = \mathbb{Q}(\vartheta, i\sqrt{m})$ with discriminant D_K and ring of integers $\mathbb{Z}_K$. Let $M = \mathbb{Q}(i\sqrt{m})$ and $L = \mathbb{Q}(\vartheta)$ be the subfields of K.

Denote by $\alpha^{(j)}$ ($1 \le j \le 6$) the conjugates of a primitive integer α in K. We have

$$I(\alpha) = \frac{\left| \prod_{1 \le j < k \le 6} \left(\alpha^{(j)} - \alpha^{(k)} \right) \right|}{\sqrt{|D_K|}}. \tag{8.9}$$

Our purpose is to find all integers $\alpha \in \mathbb{Z}_K$ with given (non–zero) index $I_0 \in \mathbb{Z}$, that is we consider the solutions of the equation

$$I(\alpha) = I_0 \quad \text{in} \quad \alpha \in \mathbb{Z}_K. \tag{8.10}$$

Set

$$\omega = \begin{cases} (1 + i\sqrt{m})/2 & \text{if } -m \equiv 1 \pmod 4, \\ i\sqrt{m} & \text{if } -m \equiv 2, 3 \pmod 4. \end{cases} \tag{8.11}$$

We represent any $\alpha \in \mathbb{Z}_K$ in the form

$$\alpha = \frac{x_0 + x_1\vartheta + x_2\vartheta^2 + y_0\omega + y_1\omega\vartheta + y_2\omega\vartheta^2}{g} \qquad (8.12)$$

with $x_0, x_1, x_2, y_0, y_1, y_2 \in \mathbb{Z}$ and with a fixed denominator $g \in \mathbb{Z}$. (In the present situation ϑ is a cubic algebraic element, hence the denominator is needed contrary to (8.1)).

Set $\mathcal{O} = \mathbb{Z}[1, \vartheta, \vartheta^2, \omega, \omega\vartheta, \omega\vartheta^2]$ and denote by $D_{\mathcal{O}}$ the discriminant of this order. It is easily seen (cf. [Ga95]) that

$$I_1 = \frac{g^{15} I_0 \sqrt{|D_K|}}{\sqrt{|D_{\mathcal{O}}|}}$$

is an integer. Denote by $\vartheta = \vartheta_1, \vartheta_2, \vartheta_3$ the conjugates of ϑ over L and let $\rho = -\vartheta_2 - \vartheta_3 \in L$. In our case Theorem 8.1.1 gets the following formulation:

Theorem 8.1.2 *If $\alpha \in \mathbb{Z}_K$ is a solution of equation (8.10) and $x_0, x_1, x_2, y_0, y_1, y_2 \in \mathbb{Z}$ are the coefficients of α in the representation (8.12), then*

$$N_{K/M}((x_1 + \omega y_1) - \rho(x_2 + \omega y_2)) = \mu, \qquad (8.13)$$
$$N_{L/\mathbb{Q}}(y_0 + y_1\vartheta + y_2\vartheta^2) = d, \qquad (8.14)$$

where $\mu \in \mathbb{Z}_M, d \in \mathbb{Z}$ such that $d \cdot N_{M/\mathbb{Q}}(\mu)$ divides I_1.

The proof of this theorem is similar to that of Theorem 8.1.1.

The (finitely many) possible values for μ, d can be determined easily. For all these pairs we have to solve the system of equations (8.13), (8.14). In the following (cf. Theorem 8.1.3) we give a simple algorithm to determine the solutions x_1, x_2, y_1, y_2 of equation (8.13). Once y_1 and y_2 are known, (8.14) is a cubic polynomial equation in y_0.

Let us fix an arbitrary solution α of (8.10) and denote by $x_0, x_1, x_2, y_0, y_1, y_2$ the coefficients of α in the representation (8.12). Set $X = x_1 + \omega y_1, Y = x_2 + \omega y_2$. Then equation (8.13) can be written as

$$N_{K/M}(X - \rho Y) = \mu \quad \text{in } X, Y \in \mathbb{Z}_M. \qquad (8.15)$$

Under our assumptions on the field K, the resolution of this equation can be simplified further, and can be reduced to the resolution of cubic Thue inequalities over $\mathbb{Z}$. Denote by $\rho = \rho_1, \rho_2, \rho_3$ the conjugates of ρ over L. Choose the indices $\{r, s, t\} = \{1, 2, 3\}$ according to

$$|X - \rho_r Y| \leq |X - \rho_s Y| \leq |X - \rho_t Y|. \qquad (8.16)$$

Set

$$
c_m = \begin{cases} 2 & \text{if } -m \equiv 1 \ (\text{mod } 4), \\ 1 & \text{if } -m \equiv 2, 3 \ (\text{mod } 4), \end{cases}
$$

$$
c_1 = 9c_m^3 |\mu|,
$$

$$
c_2 = \min(|\rho_r - \rho_s|, |\rho_r - \rho_t|),
$$

$$
c_3 = |\rho_r - \rho_s| \cdot |\rho_r - \rho_t|,
$$

$$
c_4 = \max \left\{ \frac{2|\mu|^{1/3}}{c_2}, \frac{4c_m |\mu|}{c_3 \sqrt{m}} \right\},
$$

$$
c_5 = \left(\frac{8|\mu|}{c_2 c_3} \right)^{1/3}.
$$

Finally put

$$
F(x, y) = \prod_{j=1}^{3} (x - \rho_j y) \in \mathbb{Z}[x, y].
$$

Under these assumptions we have the following theorem:

Theorem 8.1.3 *Let* $X = x_1 + \omega y_1$, $Y = x_2 + \omega y_2 \in \mathbb{Z}_M$ *be a solution of (8.15) according to (8.16). Suppose* $|Y| > c_4$. *We have*

$$
x_1 y_2 = x_2 y_1. \tag{8.17}
$$

Further, in case $-m \equiv 1 \pmod 4$

$$
\begin{aligned}
&\text{if} \ \ |2x_2 + y_2| \geq 2c_5, &&\text{then} \ \ |F(2x_1 + y_1, 2x_2 + y_2)| \leq c_1; \\
&\text{if} \ \ |y_2| \geq 2c_5/\sqrt{m}, &&\text{then} \ \ |F(y_1, y_2)| \leq c_1/(\sqrt{m})^3;
\end{aligned}
$$

and in case $-m \equiv 2, 3 \pmod 4$

$$
\begin{aligned}
&\text{if} \ \ |x_2| \geq c_5, &&\text{then} \ \ |F(x_1, x_2)| \leq c_1; \\
&\text{if} \ \ |y_2| \geq c_5/\sqrt{m}, &&\text{then} \ \ |F(y_1, y_2)| \leq c_1/(\sqrt{m})^3.
\end{aligned}
$$

Theorem 8.1.3 implies that the resolution of equation (8.15) can be reduced to the resolution of Thue inequalities of type

$$
|F(a, b)| \leq c \quad \text{in} \ \ a, b \in \mathbb{Z} \tag{8.18}
$$

and to testing some small values of the variables. One has to solve this inequality only once, with right side $c = c_1$; the solutions of the other inequality with right side $c = c_1/(\sqrt{m})^3$ are contained in the set of solutions of that inequality. Note that $c_1/(\sqrt{m})^3$ is often very small and for $c < 1$ inequality (8.18) has only the trivial solution $a = b = 0$. For the resolution of Thue inequalities see Section 3.1.3. (Section 3.1.3 describes how to find "small" solutions of Thue inequalities. If we need "all" solutions, then we need to solve the corresponding Thue equations with all possible right-hand sides.) For testing small values of x_2, y_2 observe that for fixed x_2, y_2 equation (8.13) is a cubic polynomial equation in $X = x_1 + \omega y_1$ allowing us to determine x_1 and y_1 by separating the real and imaginary parts of X.

To prove our theorem we shall need the following lemma:

Lemma 8.1.1 *Let c_0 be a given positive constant, a, b integers, $b \neq 0$. If*

$$\left| \rho_r - \frac{a}{b} \right| \leq \frac{c_0}{b^3} \quad \text{and} \quad |b| \geq \left(\frac{2c_0}{c_2} \right)^{1/3}, \tag{8.19}$$

then

$$|F(a,b)| \leq \frac{9}{4} c_0 c_3. \tag{8.20}$$

Proof of Lemma 8.1.1. By combining (8.19) for $j = s, t$ we have

$$\left| \rho_j - \frac{a}{b} \right| \leq |\rho_j - \rho_r| + \left| \rho_r - \frac{a}{b} \right| \leq \frac{3}{2} |\rho_j - \rho_r|.$$

Now this inequality together with (8.19) implies

$$\left| \left(\rho_r - \frac{a}{b} \right) \left(\rho_s - \frac{a}{b} \right) \left(\rho_t - \frac{a}{b} \right) \right| \leq \frac{9 c_0 c_3}{4 b^3},$$

from which we obtain (8.20).   □

Proof of Theorem 8.1.3. Equations (8.15) and (8.16) imply

$$|X - \rho_r Y| \leq |\mu|^{1/3}. \tag{8.21}$$

For $j = s, t$ by $|Y| > c_4$ and (8.21) we obtain

$$|X - \rho_j Y| = |(\rho_r - \rho_j)Y + (X - \rho_r Y)|$$

$$\geq |\rho_j - \rho_r| \cdot |Y| - |\mu|^{1/3} \geq \frac{1}{2} |\rho_j - \rho_r| \cdot |Y|. \tag{8.22}$$

Combining (8.21), (8.22), equation (8.15) implies

$$|X - \rho_r Y| \leq \frac{c_6}{|Y|^2} \tag{8.23}$$

with $c_6 = 4|\mu|/c_3$. Dividing by $|Y|$, $(|Y| > c_4)$ we get

$$\left| \rho_r - \frac{X}{Y} \right| \leq \frac{c_6}{|Y|^3}, \tag{8.24}$$

whence

$$\left| \rho_r - \frac{X\overline{Y}}{|Y|^2} \right| \le \frac{c_6}{|Y|^3},$$

where $\overline{Y}$ is the complex conjugate of Y, being the same as the conjugate of Y over M. Separating the imaginary part, the above inequality implies

$$\mathrm{Im}(X\overline{Y}) \le \frac{c_6}{|Y|}. \tag{8.25}$$

The smallest possible imaginary part (in absolute value) of a non-zero integer in M is $\sqrt{m}/2$ if $-m \equiv 1 \pmod 4$ and $\sqrt{m}$ if $-m \equiv 2, 3 \pmod 4$. The assumption $|Y| > c_4$ in both cases implies $\mathrm{Im}(X\overline{Y}) = 0$, which is equivalent to (8.17).

In case $-m \equiv 1 \pmod 4$ equation (8.23) has the form

$$\left| \left(x_1 + \frac{1+i\sqrt{m}}{2} y_1 \right) - \rho_r \left(x_2 + \frac{1+i\sqrt{m}}{2} y_2 \right) \right| \le \frac{c_6}{\left(x_2 + \frac{y_2}{2} \right)^2 + \frac{m}{4} y_2^2},$$

whence

$$\left| \left[\left(x_1 + \frac{y_1}{2} \right) - \rho_r \left(x_2 + \frac{y_2}{2} \right) \right] + \frac{i\sqrt{m}}{2} [y_1 - \rho_r y_2] \right| \le \frac{c_6}{\left(x_2 + \frac{y_2}{2} \right)^2 + \frac{m}{4} y_2^2}.$$

Separating the real and imaginary parts, in case $2x_2 + y_2 \ne 0$ the above equation implies

$$\left| \rho_r - \frac{2x_1 + y_1}{2x_2 + y_2} \right| \le \frac{8c_6}{(2x_2 + y_2)^3}, \tag{8.26}$$

and in case $y_2 \ne 0$ we obtain

$$\left| \rho_r - \frac{y_1}{y_2} \right| \le \frac{8c_6}{\sqrt{m}^3} \frac{1}{y_2^3}. \tag{8.27}$$

Similarly, in case $-m \equiv 2, 3 \pmod 4$ equation (8.23) gets the form

$$\left| (x_1 + i\sqrt{m} y_1) - \rho_r(x_2 + i\sqrt{m} y_2) \right| \le \frac{c_6}{x_2^2 + my_2^2},$$

whence

$$\left| (x_1 - \rho_r x_2) + i\sqrt{m}(y_1 - \rho_r y_2) \right| \le \frac{c_6}{x_2^2 + my_2^2}.$$

Again separating the real and imaginary parts, in case $x_2 \ne 0$ we get

$$\left| \rho_r - \frac{x_1}{x_2} \right| \le \frac{c_6}{x_2^3} \tag{8.28}$$

and in case $y_2 \ne 0$ we conclude

$$\left| \rho_r - \frac{y_1}{y_2} \right| \le \frac{c_6}{\sqrt{m}^3} \frac{1}{y_2^3}. \tag{8.29}$$

Applying Lemma 8.1.1 to (8.26), (8.27), (8.28), (8.29) we obtain the assertion of Theorem 8.1.3.　□

8.1.5 Parametric families of sextic fields with imaginary quadratic and real cubic subfields

The power of the results of Section 8.1.4 is shown by the following application to an infinite family of fields with two parameters (cf. I. Gaál [Ga95]). Namely, we compose Shanks' simplest cubics (cf. D. Shanks, [Sh74], Section 5.2) with imaginary quadratic number fields.

Let a be a natural number, let ϑ be a root of

$$f_a(x) = x^3 - ax^2 - (a+3)x - 1 \tag{8.30}$$

and let m be a square-free positive integer. Consider the two-parametric family $K = \mathbb{Q}(\vartheta, i\sqrt{m})$ of totally complex cyclic sextic fields. Define ω as in (8.11) and set $\mathcal{O} = \mathbb{Z}[1, \vartheta, \vartheta^2, \omega, \omega\vartheta, \omega\vartheta^2]$ with discriminant $D_{\mathcal{O}}$ as before. We also use $L = \mathbb{Q}(\vartheta)$ and $M = \mathbb{Q}(i\sqrt{m})$ as in the preceding section. Unfortunately it seems to be difficult to describe an integral basis of K. On the other hand, the order $\mathcal{O}$ is very often the maximal order of K. Hence we restrict ourselves to considering power integral bases in $\mathcal{O}$. Put

$$m_0 = \begin{cases} 19 & \text{if} \quad -m \equiv 1 \pmod 4, \\ 5 & \text{if} \quad -m \equiv 2, 3 \pmod 4. \end{cases}$$

Theorem 8.1.4 *Assume that $a \geq 3$ and $m \geq m_0$. Then the order $\mathcal{O}$ has no power integral bases.*

Proof of Theorem 8.1.4. In our situation we apply Theorems 8.1.2 and 8.1.3 with $I_0 = I_1 = 1$. Denote by $\vartheta = \vartheta_1 < \vartheta_2 < \vartheta_3$ the roots of (8.30) and put $\rho = -\vartheta_2 - \vartheta_3$, that is $\rho_j = \vartheta_j - a$ for $1 \leq j \leq 3$. We have (cf. E. Thomas [Tho90])

$$\vartheta_3 \geq 1 + a, \quad \vartheta_2 = -\frac{1}{1 + \vartheta_3}, \quad \vartheta_1 = -1 - \frac{1}{\vartheta_3},$$

thus

$$\rho_2 - \rho_1 = 1 + \frac{1}{\vartheta_3(\vartheta_3 + 1)}, \quad \rho_3 - \rho_2 = \vartheta_3 + \frac{1}{\vartheta_3 + 1}.$$

Hence, if $a \geq 3$, then for any choice of the indices r, s, t the constants of Theorem 8.1.3 satisfy

$$c_2 \geq 1, \quad c_3 \geq a + 1, \quad c_4 \leq 2, \quad c_5 \leq \frac{2}{(a+1)^{1/3}}.$$

A. Consider now the solutions with $|Y| \leq c_4$. In view of $m \geq m_0$ it implies $|x_2| \leq 2$ and $y_2 = 0$ both for $-m \equiv 1 \pmod 4$ and for $-m \equiv 2, 3 \pmod 4$. For $y_2 = 0$ equation (8.14) reduces to

$$N_{L/\mathbb{Q}}(y_0 + \vartheta y_1) = \pm 1,$$

whence by Lemma 5.2.1 of E. Thomas and M. Mignotte we have $(y_0, y_1) = (\pm 1, 0), (0, \pm 1), (\pm 1, \pm 1)$.

A1. For $y_0 = \pm 1$, $y_1 = 0$, $y_2 = 0$ equation (8.13) becomes

$$N_{L/\mathbb{Q}}(x_1 - \rho x_2) = \pm 1.$$

Using again Lemma 5.2.1 we obtain $(x_1, x_2) = (1, 0)$, $(a, -1)$, $(a + 1, -1)$ and their negatives. Note that the index form equation corresponding to the given basis of the order $\mathcal{O}$ has three factors with coefficients in $\mathbb{Z}$. Two of them imply equations (8.13) and (8.14) of Theorem 8.1.2, the third factor is of degree 6 (the right-hand sides of all three equations are ± 1). For $y_0 = \pm 1$, $y_1 = 0$, $y_2 = 0$ and the above (x_1, x_2) we tested the third equation implied by the third factor of the index form. Using symmetric polynomials we calculated the coefficients of the third factor, and substituted the above three pairs for (x_1, x_2). Surprisingly in all three cases we obtained the same result, namely

$$(m + 1)a^4 + (6m + 6)a^3 + (2m^2 + 27m + 27)a^2$$
$$+ (6m^2 + 54m + 54)a + (m^3 + 18m^2 + 81m + 81)$$

for $-m \equiv 1 \pmod 4$ and

$$(4m + 1)a^4 + (24m + 6)a^3 + (32m^2 + 108m + 27)a^2$$
$$+ (96m^2 + 216m + 54)a + (64m^3 + 288m^2 + 324m + 81)$$

for $-m \equiv 2, 3 \pmod 4$. These expressions can not attain the values ± 1.

A2. For $y_1 = \varepsilon = \pm 1$, $y_2 = 0$ equation (8.13) takes the form

$$N_{K/M}((x_1 + \varepsilon \omega) - \rho x_2) = \mu, \tag{8.31}$$

where $\mu \in \mathbb{Z}_M$ with $N_{M/\mathbb{Q}}(\mu) = |\mu|^2 = 1$.

If $x_2 = 0$, then it implies $|x_1 + \varepsilon \omega| = 1$, which is impossible for $m \geq m_0$. Consider now the cases $x_2 = \pm 1, \pm 2$. Then (8.31) can be written in the form

$$\prod_{j=1}^{3} \left(\frac{x_1 + \varepsilon \omega}{x_2} - \rho_j \right) = \frac{\mu}{x_2^3}. \tag{8.32}$$

Denote the left side of (8.32) by g_1 for $-m \equiv 1 \pmod 4$. We have

$$\mathrm{Im}(g_1) = \frac{\sqrt{m}}{8x_2^3} G_1,$$

where G_1 is a polynomial in x_1, x_2, a, m with integer coefficients. The discriminant of G_1 with respect to x_1 is

$$D_1 = 2^4 x_2^2 (3m + 4x_2^2 a^2 + 12x_2^2 a + 36x_2^2) > 0,$$

whence G_1 is a non–zero rational integer. The minimum of $|G_1|$ as a quadratic polynomial in x_1 is $D_1/(48x_2^2)$ whence $|G_1| \geq 16$. We have

$$\frac{2\sqrt{m}}{|x_2|^3} \leq \left| \frac{\sqrt{m}}{8x_2^3} G_1 \right| = |\mathrm{Im}(g_1)| \leq |g_1| = \left| \frac{\mu}{x_2^3} \right| = \frac{1}{|x_2|^3},$$

that is $\sqrt{m} \leq 1/2$ which is impossible.

Similarly, denote the left side of (8.32) by g_2 for $-m \equiv 2, 3 \pmod 4$. Then

$$\mathrm{Im}(g_2) = \frac{\sqrt{m}}{x_2^3} G_2,$$

where G_2 is a polynomial in x_1, x_2, a, m with integer coefficients. The discriminant of G_2 with respect to x_1 is

$$D_2 = 2^2 x_2^2 (3m + x_2^2 a^2 + 3x_2^2 a + 9x_2^2) > 0,$$

whence G_2 is a non–zero rational integer. Hence

$$\left| \frac{\sqrt{m}}{x_2^3} G_2 \right| = |\mathrm{Im}(g_2)| \leq |g_2| = \left| \frac{\mu}{x_2^3} \right| = \frac{1}{|x_2|^3},$$

that is $\sqrt{m} \leq 1$, again impossible.

B. We still have to check the solutions with $|Y| > c_4$. In case $-m \equiv 1 \pmod 4$ by Theorem 8.1.3 either $|y_2| < 2c_5/\sqrt{m} < 1$, that is $y_2 = 0$, or in the opposite case if $|y_2| \geq 2c_5/\sqrt{m} > 0$, then

$$|F(y_1, y_2)| < \frac{9c_m^3}{\sqrt{m}^3} < 1$$

implying $y_1 = y_2 = 0$ (for $m \geq m_0$) and contradiction with $|y_2| > 0$. Hence $y_2 = 0$. We get the same result for $-m \equiv 2, 3 \pmod 4$ with $2c_5/\sqrt{m}$ replaced by $c_5/\sqrt{m}$.

Now equation (8.14) reduces to

$$N_{L/\mathbb{Q}}(y_0 + \vartheta y_1) = \pm 1,$$

whence by Lemma 5.2.1 we conclude $(y_0, y_1) = (\pm 1, 0), (0, \pm 1), (\pm 1, \pm 1)$.

B1. The case $y_0 = \pm 1$, $y_1 = 0$, $y_2 = 0$ was already considered in A1.

B2. If $y_1 = \pm 1$, $y_2 = 0$, then by (8.7) we get $x_2 = 0$. In this case equation (8.13) reduces to $|x_1 + \varepsilon \omega| = 1$ being impossible for $m \geq m_0$ (cf. A2).   □

Another infinite parametric family of sextic fields with an imaginary quadratic and a real cubic subfield was considered by P. Olajos [Ola01]. Here we only state the result, the proof is based on Theorem 8.1.3 and is roughly similar to the above arguments:

Theorem 8.1.5 *Let $a \geq 3$ be an integer, ϑ a root of the polynomial $f_a(x) = x^3 - ax^2 - (a+1)x - 1$. Let $m \geq 1$ be a square-free integer and define ω as in (8.11). Set*

$$m_0 = \begin{cases} 36 & \text{if } -m \equiv 1 \pmod 4, \\ 9 & \text{if } -m \equiv 2, 3 \pmod 4. \end{cases}$$

If $a \geq 7$ and $m \geq m_0$, then the order $\mathcal{O} = \mathbb{Z}[1, \vartheta, \vartheta^2, \omega, \omega\vartheta, \omega\vartheta^2]$ of the field $K = \mathbb{Q}(\vartheta, i\sqrt{m})$ has no power integral basis.

The proof of Theorem 8.1.5 used also the description of all solutions of the family of Thue equations $x^3 - ax^2y - (a + 1)xy^2 - y^3 = \pm 1$ given by M. Mignotte and N. Tzanakis [MT91]. In his paper P. Olajos [Ola01] investigated also the parameters outside the region $a \geq 7$ and $m \geq m_0$ by direct computations.

8.2 Sextic fields with a cubic subfield

Recently I. Járási [Ja01] considered the resolution of index form equations in sextic fields having only a cubic subfield (and no quadratic subfields).

Let $K = M(\vartheta)$ be a sextic field having a cubic subfield $M = \mathbb{Q}(\varrho)$, generated by algebraic integers ϑ and ϱ. Denote by $\mathbb{Z}_K$ and $\mathbb{Z}_M$ the rings of integers of K and M, respectively. We shall consider here the more interesting case of totally real cubic subfields M, the case of complex cubic subfields is easier. Denote by η_1 and η_2 the fundamental units of M. The conjugates of ϱ are $\varrho^{(i)}$ ($i = 1, 2, 3$), and the roots of i-th conjugate of the quadratic relative defining polynomial of ϑ over $M^{(i)} = M(\varrho^{(i)})$ are $\vartheta^{(i)}, \overline{\vartheta^{(i)}}$ ($i = 1, 2, 3$). The conjugates corresponding to $\varrho^{(i)}$ of any $\delta \in M$ will be denoted by $\delta^{(i)}$ the conjugates of any $\gamma \in K$ corresponding to $\vartheta^{(i)}, \overline{\vartheta^{(i)}}$ will be denoted by $\gamma^{(i)}, \overline{\gamma^{(i)}}$, respectively ($i = 1, 2, 3$). For simplicity we assume that $\{1, \varrho, \varrho^2, \vartheta, \vartheta\varrho, \vartheta\varrho^2\}$ is an integral basis in K, that is any $\alpha \in \mathbb{Z}_K$ can be represented in the form

$$\alpha = x_0 + x_1\varrho + x_2\varrho^2 + y_0\vartheta + y_1\vartheta\varrho + y_2\vartheta\varrho^2 \tag{8.33}$$

with $x_0, x_1, x_2, y_0, y_1, y_2 \in \mathbb{Z}$. (Otherwise we have to use a common denominator d in the above expression and corresponding constants occur also on the right-hand sides of some of the following equations.) Set

$$X = x_0 + x_1\varrho + x_2\varrho^2, \quad Y = y_0 + y_1\varrho + y_2\varrho^2,$$

then $X, Y \in \mathbb{Z}_M$ and

$$\alpha = X + \vartheta Y. \tag{8.34}$$

For $k = 1, 2, 3$ set $(i, j) = (2, 3), (3, 1), (1, 2)$, respectively. Define

$$\beta^{(k)} = \frac{\left(\alpha^{(i)} - \alpha^{(j)}\right)\left(\overline{\alpha^{(i)}} - \overline{\alpha^{(j)}}\right)\left(\overline{\alpha^{(i)}} - \alpha^{(j)}\right)\left(\alpha^{(i)} - \overline{\alpha^{(j)}}\right)}{\left(\varrho^{(i)} - \varrho^{(j)}\right)^2}. \tag{8.35}$$

Using the actions of the Galois automorphisms I. Járási [Ja01] showed:

Theorem 8.2.1 *The element $\beta^{(k)}$ is an algebraic integer in $M^{(k)} = \mathbb{Q}(\varrho^{(k)})$ ($k = 1, 2, 3$). The element α of (8.33) generates a power integral basis in K if and only if*

$$N_{M/\mathbb{Q}}(y_0 + y_1\varrho + y_2\varrho^2) = \pm 1, \tag{8.36}$$

$$N_{M/\mathbb{Q}}(\beta) = \pm 1. \tag{8.37}$$

The above theorem makes it possible to find the generators α of power integral bases of K having small coefficients in the integral basis, say with $\max(|x_1|, |x_2|, |y_0|, |y_1|, |y_2|) < C = 10^5$ which range contains all solutions with a high probability. The bound for the solutions imply also a bound for the exponents in the representation

$$y_0 + y_1 \varrho + y_2 \varrho^2 = \pm \eta_1^{b_1} \eta_2^{b_2}. \tag{8.38}$$

The order of magnitude of the bound derived for $|b_1|, |b_2|$ is $\log C$. For each pair b_1, b_2 we can determine the corresponding y_0, y_1, y_2 and Y from equation (8.38) by taking conjugates and solving the corresponding system of linear equations. Also, for the given y_0, y_1, y_2 and $|x_1|, |x_2| < C$ we can derive an upper bound for the absolute values of

$$\beta^{(k)} = \pm \left(\eta_1^{(k)} \right)^{d_1} \left(\eta_2^{(k)} \right)^{d_2} \tag{8.39}$$

and for the exponents d_1, d_2. We have

$$\begin{aligned}
\alpha^{(i)} - \alpha^{(j)} &= X^{(i)} - X^{(j)} + \vartheta^{(i)} Y^{(i)} - \vartheta^{(j)} Y^{(j)}, \\
\overline{\alpha^{(i)}} - \overline{\alpha^{(j)}} &= X^{(i)} - X^{(j)} + \overline{\vartheta^{(i)}} Y^{(i)} - \overline{\vartheta^{(j)}} Y^{(j)}, \\
\overline{\alpha^{(i)}} - \alpha^{(j)} &= X^{(i)} - X^{(j)} + \overline{\vartheta^{(i)}} Y^{(i)} - \vartheta^{(j)} Y^{(j)}, \\
\alpha^{(i)} - \overline{\alpha^{(j)}} &= X^{(i)} - X^{(j)} + \vartheta^{(i)} Y^{(i)} - \overline{\vartheta^{(j)}} Y^{(j)},
\end{aligned} \tag{8.40}$$

hence by (8.35) for fixed Y and for each pair d_1, d_2 (8.39) gives a quartic equation for $X^{(i)} - X^{(j)}$ which makes it possible to determine x_1, x_2, as well. The element α of (8.33) is to be tested if it has index one.

Using the methods of Section 2, I. Járási is going to construct a feasible algorithm also for the complete resolution of the system of equations (8.36), (8.37).

EXAMPLE

Let ϱ be defined by $g(x) = x^3 - 5x - 1$. Then $M = \mathbb{Q}(\varrho)$ has discriminant $D_M = 11 \cdot 43$, fundamental units $\eta_1 = \varrho$, $\eta_2 = 2 + \varrho$. The relative defining polynomial of ϑ over M is $f(x) = x^2 - 10x - \varrho$, its defining polynomial over $\mathbb{Q}$ is $f_0(x) = x^6 - 30x^5 + 300x^4 - 1000x^3 - 5x^2 + 50x - 1$. The totally real sextic field $K = \mathbb{Q}(\vartheta)$ has discriminant $D_K = 2^6 \cdot 11^3 \cdot 43^2 \cdot 1409$ and integral basis $\{1, \varrho, \varrho^2, \vartheta, \vartheta\varrho, \vartheta\varrho^2\}$. The generators of power integral bases with $\max(|x_1|, |x_2|, |y_0|, |y_1|, |y_2|) < 10^5$ are

$$(x_1, x_2, y_0, y_1, y_2) = (0, -10, -5, 0, 1), (0, 0, -5, 0, 1), (0, 0, 1, 0, 0).$$

8.3   Sextic fields as composite fields

In the following we give applications of the results of Section 4.4.2 to sextic fields (cf. I. Gaál, P. Olajos and M. Pohst [GOP01]).

8.3.1 A cyclic sextic field

Consider the sextic field K generated by a root of $h(x) = x^6 - x^5 - 6x^4 + 6x^3 + 8x^2 - 8x + 1$. This is a totally real cyclic sextic field with discriminant $D_K = 453789 = 3^3 7^5$. Its cubic subfield is $L = \mathbb{Q}(\varphi)$ (with discriminant 49) where φ is a root of $f(x) = x^3 + 4x^2 + 3x - 1$. In the field L the elements $\{1, \varphi, \varphi^2\}$ form an integer basis. We have $f(x) \equiv (x + 6)^3 \pmod{7}$. The quadratic subfield is $M = \mathbb{Q}(\sqrt{21})$. The polynomial $g(x) = x^2 - x - 5$ has $\psi = (1 + \sqrt{21})/2$ as a root, and obviously $\{1, \psi\}$ is an integral basis in M. We have $g(x) \equiv (x - 1/2)^2 \pmod{7}$. Theorem 4.4.2 implies that the indices of the primitive elements of the order $\mathcal{O}_{fg} = \mathbb{Z}[1, \varphi, \varphi^2, \psi, \varphi\psi, \varphi^2\psi]$ are all divisible by 7, hence it has no power integral basis. Note that the maximal order of the same sextic field does have power integral basis, cf. Table 11.3.1.

8.3.2 A non-cyclic sextic field

Consider the sextic field K generated by a root of $h(x) = x^6 - 12190x^4 + 256565x^2 - 12167$. This is a totally real sextic field with Galois group D_6, discriminant $D_K = 2^6 17^2 23^3 647^2$. Its cubic subfield is $L = \mathbb{Q}(\varphi)$ (with discriminant $252977 = 17 \cdot 23 \cdot 647$ and Galois group S_3) where φ is a root of $f(x) = x^3 - 22x^2 - 23x - 1$. In the field L the elements $\{1, \varphi, \varphi^2\}$ form an integer basis. We have $f(x) \equiv (x + 15)(x + 16)^2 \pmod{23}$. The quadratic subfield is $M = \mathbb{Q}(\sqrt{23})$. The polynomial $g(x) = x^2 - 23$ has $\psi = \sqrt{23}$ as a root, and obviously $\{1, \psi\}$ is an integral basis in M. We have $g(x) \equiv x^2 \pmod{23}$. Theorem 4.4.2 implies that the indices of the primitive elements of the order $\mathcal{O}_{fg} = \mathbb{Z}[1, \varphi, \varphi^2, \psi, \varphi\psi, \varphi^2\psi]$ are all divisible by 23, hence it has no power integral basis.

8.3.3 The parametric family of simplest sextic fields

Assume $3 \nmid t$, $t \neq -8, -5$. Let us consider the family of sextic fields K_t generated by a root β of the polynomial

$$h_t(x) = x^6 - 2tx^5 - (5t + 15)x^4 - 20x^3 + 5tx^2 + (2t + 6)x + 1.$$

This family of fields is called the *simplest sextic fields*, having a couple of nice properties, detailed in G. Lettl, A. Pethő and P. Voutier [LPV98]. Among others these fields are totally real cyclic fields. Let $q = t^2 + 3t + 9$. We have $d(h_t) = 6^6 q^5$. Note that $h_t(x) \equiv (x - t/3)^6 \pmod{q}$ (the family of Lehmer's quintics have a similar property, cf. Section 7.2).

The cubic subfield L_t of K_t is generated by a root φ of $f_t = x^3 - tx^2 - (t+3)x - 1$ with $d(f_t) = q^2$. These are the "simplest cubic fields" of D. Shanks [Sh74], totally real, cyclic. Note that $f_t(x) \equiv (x - t/3)^3 \pmod{q}$.

The quadratic subfield of K_t is $M_t = \mathbb{Q}(\sqrt{q})$.

If $q \equiv 2, 3 \pmod{4}$, then set $g_t(x) = x^2 - q$ with $d(g_t) = 4q$ and with a root $\psi = \sqrt{q}$. In this case $g_t(x) \equiv x^2 \pmod{q}$.

If $q \equiv 1 \pmod{4}$, then set $g_t(x) = x^2 - x - (q-1)/4$ with $d(g_t) = q$ and with a root $\psi = (1 + \sqrt{q})/2$. In this case $g_t(x) \equiv (x - 1/2)^2 \pmod{q}$.

In both cases $\{1, \psi\}$ is an integer basis of M_t.

Consider now the order $\mathcal{O}_{fg} = \mathbb{Z}[1, \varphi, \varphi^2, \psi, \varphi\psi, \varphi^2\psi]$. By Theorem 4.4.2 the indices of the primitive elements of $\mathcal{O}_{fg}$ are all divisible by q, hence $\mathcal{O}_{fg}$ has no power integral bases.

9

Relative Power Integral Bases

In this chapter we open a new horizon by considering the problem of power integral bases in relative extensions instead of over $\mathbb{Q}$ as in the classical situation. The algorithms for determining generators of relative power integral bases will be applied for finding generators of integral bases in higher degree fields having subfields. It is easy to see that if an element generates a power integral basis, then it also generates a relative power integral basis over a subfield. Thus, for example the algorithm for relative quartic extensions described in Section 9.3 will be used in octic fields with a quadratic subfield in Section 10.1.

The case of relative cubic extensions considered in Section 9.2 is again easy in theory, since we just have to solve a cubic relative Thue equation. To make this case more interesting, we tried to go as far as possible and considered (totally real) relative cubic extensions of quintic and even sextic fields. We need to solve unit equations with 10 and 12 unknown exponents, respectively. This is certainly the limit of applicability of our procedures, especially of Wildanger's enumeration method.

9.1 Basic concepts

Let M be a field of degree m and let $K = M(\xi)$ be an extension of M of degree n with an algebraic integer ξ. Denote by $\mathbb{Z}_M, \mathbb{Z}_K$ the rings of integers of M, K, by D_K, D_M the discriminants of M, K, respectively. We usually represent the integers α of K in the form

$$\alpha = \frac{1}{d}\left(X_0 + X_1\xi + \cdots + X_{n-1}\xi^{n-1}\right) \tag{9.1}$$

with $X_0, X_1, \ldots, X_{n-1} \in \mathbb{Z}_M$ and with a common denominator $d \in \mathbb{Z}$.

If we investigate power integral bases in K, then the index of a primitive integer $\alpha \in K$ factorizes as

$$I(\alpha) = (\mathbb{Z}_K^+ : \mathbb{Z}^+[\alpha]) = (\mathbb{Z}_K^+ : \mathbb{Z}_M^+[\alpha]) \cdot (\mathbb{Z}_M^+[\alpha] : \mathbb{Z}^+[\alpha]), \tag{9.2}$$

where the additive groups of the corresponding modules are involved. The *relative index* of $\alpha \in \mathbb{Z}_K$ (such that $K = M(\alpha)$) with respect to the extension K/M is

$$I_{K/M}(\alpha) = (\mathbb{Z}_K^+ : \mathbb{Z}_M^+[\alpha]).$$

By the above remark, if α generates a power integral basis in K, then it must have relative index 1 over M.

In the following we also use $\mathcal{O} = \mathbb{Z}_M[\xi]$ and $i_0 = (\mathbb{Z}_K^+ : \mathcal{O}^+)$. For $\gamma \in K$ we denote by $\gamma^{(ij)}$, $j = 1, \ldots, n$ the conjugates of γ so that $K^{(ij)}$ are the images of those embeddings of K which leave the conjugate field $M^{(i)}$ of M elementwise fixed ($i = 1, \ldots, m$).

9.2   Relative cubic extensions

Power integral bases in cubic relative extensions were considered by I. Gaál [Ga01]. Let M be a field of degree m and let $K = M(\xi)$ be a cubic extension of M, with an algebraic integer ξ.

According to (9.1) any $\alpha \in \mathbb{Z}_K$ can be written in the form

$$\alpha = \frac{X_0 + X_1\xi + X_2\xi^2}{d} \tag{9.3}$$

with $X_0, X_1, X_2 \in \mathbb{Z}_M$. We have

$$\frac{d^{3m} \cdot (\mathbb{Z}_K^+ : \mathbb{Z}_M^+[\alpha])}{i_0}$$

$$= \prod_{i=1}^{m} \prod_{1 \le j_1 < j_2 \le 3} \left| \frac{d\left(\alpha^{(ij_1)} - \alpha^{(ij_2)}\right)}{\xi^{(ij_1)} - \xi^{(ij_2)}} \right|$$

$$= \prod_{i=1}^{m} \prod_{1 \le j_1 < j_2 \le 3} \left| X_1^{(i)} + \left(\xi^{(ij_1)} + \xi^{(ij_2)}\right) X_2^{(i)} \right|. \tag{9.4}$$

Denote by β the coefficient of the quadratic term of the cubic relative minimal polynomial of ξ over M, that is $\beta^{(i)} = -\xi^{(i1)} - \xi^{(i2)} - \xi^{(i3)}$, $i = 1, \ldots, m$. Then the above product can be written in the form

$$\prod_{i=1}^{m} \prod_{j=1}^{3} \left| X_1^{(i)} - \left(\beta^{(i)} + \xi^{(ij)}\right) X_2^{(i)} \right|,$$

hence by setting $\rho = \beta + \xi$, (9.4) implies

$$N_{M/\mathbb{Q}} \left(N_{K/M} \left(X_1 - \rho X_2 \right) \right) = \frac{d^{3m} \cdot (\mathbb{Z}_K^+ : \mathbb{Z}_M^+[\alpha])}{i_0}. \tag{9.5}$$

Summarizing, by (9.5) we have:

Theorem 9.2.1 *The element α of (9.3) generates a power integral basis $\{1, \alpha, \alpha^2\}$ of $\mathbb{Z}_K$ over $\mathbb{Z}_M$ if and only if $(X_1, X_2) \in \mathbb{Z}_M^2$ is a solution of the relative Thue equation*

$$N_{M/\mathbb{Q}} \left(N_{K/M} \left(X_1 - \rho X_2 \right) \right) = \frac{d^{3m}}{i_0}. \tag{9.6}$$

This relative Thue equation can be solved by the method of Section 3.3.

Our computations show that equation (9.6) is feasible to solve even for quintic or sextic base fields M and totally real K. Note that in [Ga01] it was the first case when cubic relative Thue equations were solved over quintic and sextic fields. For the resolution of these relative Thue equations we have to solve unit equations in $r = 10$ resp. $r = 12$ unknown exponents. The examples are very interesting from a computational point of view, hence we give a somewhat more detailed description of the computations.

9.2.1 Example 1. Cubic extension of a quintic field

Let $M = \mathbb{Q}(\mu)$ where μ has minimal polynomial $f(x) = x^5 - 5x^3 + x^2 + 3x - 1$. This totally real quintic field has integral basis $\{1, \mu, \mu^2, \mu^3, \mu^4\}$ and discriminant $D_M = 24217 = 61 \cdot 397$.

Consider now the cubic field $L = \mathbb{Q}(\xi)$ where ξ has minimal polynomial $g(x) = x^3 - x^2 - 4x + 3$. This totally real cubic field has integral basis $\{1, \xi, \xi^2\}$ and discriminant $D_L = 257$.

The totally real composite field $K = LM = M(\xi)$ is of degree 15 generated by $\mu\xi$ over $\mathbb{Q}$ with minimal polynomial

$$\begin{aligned} h(x) \ = \ & x^{15} - 45x^{13} + 4x^{12} + 661x^{11} - 76x^{10} - 3763x^9 + 599x^8 \\ & + 9774x^7 - 1911x^6 - 11785x^5 + 2565x^4 + 5877x^3 \\ & - 1323x^2 - 972x + 243. \end{aligned}$$

Since $(D_M, D_L) = 1$, the elements $\{\mu^i \xi^j : i = 0, \ldots, 4, j = 0, 1, 2\}$ form an integral basis of K. We have

$$D_K = 159230640476291879672088841 = 61^3 \cdot 397^3 \cdot 257^5.$$

Hence $d = 1$ in (9.3) and $i_0 = (\mathbb{Z}_K^+ : \mathcal{O}^+) = 1$.

The fundamental units of K and M were computed by using KASH [DF97].

The set of fundamental units of M formed a subset of the set of fundamental units of K. Hence we had $r = 10$ relative units.

In the unit equation (cf. (2.5), (3.29)) we had $r = 10$ unknown exponents. Baker's method gave $A < 10^{86}$ for the exponents of this unit equation. The reduction algorithm (Lemma 2.2.2) was used with 11 terms in the linear form, according to Table 9.1. Using the notation of Lemma 2.2.2 in each step A_0 denotes the original bound for A, H is the constant playing an important role in the corresponding lattice, A_1 is the reduced bound for A. Table 9.1 includes the number of digits (precision) used for the computation and the execution time of the reduction step.

Table 9.1:

Step	A_0	H	A_1	Digits	CPUtime
1.	10^{86}	10^{900}	1962	1500	180 min
2.	1962	10^{50}	113	150	3 min
3.	113	10^{40}	92	150	3 min
4.	92	10^{35}	80	150	3 min
5.	80	10^{33}	75	150	3 min

The final bound $A < 75$ implied the bound $S_0 = 10^{1518}$ in (2.9).

In the enumeration procedure (cf. Section 2.3) we had 15 ellipsoids in 10 variables. The enumeration of the integer points of the ellipsoids were performed in several steps, according to Table 9.2. Using the notation of Section 2.3 the table includes S, s the number of digits used, the number of tuples enumerated in the 15 ellipsoids together and the execution time. The last line corresponds to the ellipsoid (2.17).

The exponent tuples were tested if there are solutions corresponding to them. The element $\alpha \in \mathbb{Z}_K$ generates a relative power integral basis of K over M if and only if it is of the form $\alpha = X_0 + \varepsilon(X_1\xi + X_2\xi^2)$ with arbitrary $X_0 \in \mathbb{Z}_M$, an arbitrary unit ε in M and $X_1 = x_0 + x_1\mu + x_2\mu^2 + x_3\mu^3 + x_4\mu^4$, $X_2 = y_0 + y_1\mu + y_2\mu^2 + y_3\mu^3 + y_4\mu^4$, whose coordinates are listed in Table 9.3.

9.2.2   Example 2. Cubic extension of a sextic field

Let $M = \mathbb{Q}(\mu)$ where μ has minimal polynomial $f(x) = x^6 - 5x^5 + 2x^4 + 18x^3 - 11x^2 - 19x + 1$. This totally real sextic field has integral basis $\{1, \mu, \mu^2, \mu^3, \mu^4, \mu^5\}$ and discriminant $D_M = 592661$ (prime).

Consider now the cubic field $L = \mathbb{Q}(\xi)$ where ξ has minimal polynomial $g(x) = x^3 - x^2 - 4x + 3$. This is the same totally real cubic field as in Example 1. The field L has integral basis $\{1, \xi, \xi^2\}$ and discriminant $D_L = 257$.

The totally real composite field $K = LM = M(\xi)$ is of degree 18 generated by $\mu\xi$ over $\mathbb{Q}$ with minimal polynomial

$$
\begin{aligned}
h(x) = {} & x^{18} - 5x^{17} - 82x^{16} + 397x^{15} + 2501x^{14} - 11919x^{13} \\
& - 34100x^{12} + 169532x^{11} + 187998x^{10} - 1174096x^9 \\
& - 154240x^8 + 3624928x^7 - 1182695x^6 - 4239690x^5 \\
& + 1472949x^4 + 1786860x^3 - 107325x^2 - 18468x + 729.
\end{aligned}
$$

Table 9.2:

Step	S	s	Digits	tuples	CPUtime
1.	10^{1518}	10^{50}	200	0	7.0 min
2.	10^{50}	10^{20}	70	0	2.7 min
3.	10^{20}	10^{12}	50	28	1.9 min
4.	10^{12}	10^{10}	50	30	1.5 min
5.	10^{10}	10^{8}	50	617	1.5 min
6.	10^{8}	10^{7}	50	899	1.6 min
7.	10^{7}	10^{6}	50	2629	2.0 min
8.	10^{6}	10^{5}	50	6513	2.7 min
9.	10^{5}	$10^{4.5}$	50	4016	2.1 min
10.	$10^{4.5}$	10^{4}	50	4974	2.2 min
11.	10000	6000	40	2848	1.5 min
12.	6000	3000	40	3390	1.6 min
13.	3000	1500	40	3192	1.5 min
14.	1500	1000	40	2132	1.3 min
15.	1000	500	40	2554	1.3 min
16.	500	250	40	2007	1.2 min
17.	250	150	40	1137	0.9 min
18.	150	100	40	722	0.8 min
19.	100	50	40	715	0.9 min
20.	50	25	40	345	0.7 min
21.	25	12	40	136	0.5 min
22.	12	6	40	45	0.4 min
23.	6	3	40	30	0.3 min
24.	3		40	2	0.2 min

Since $(D_M, D_L) = 1$ the elements $\{\mu^i \xi^j : \quad i = 0, \ldots, 5, \ j = 0, 1, 2\}$ form an integral basis of K. We have

$$D_K = 59981564379238299956091922221869 = 257^6 \cdot 592661^3.$$

Hence $d = 1$ in (9.3) and $i_0 = (\mathbb{Z}_K^+ : \mathcal{O}^+) = 1$.

The fundamental units of K and M were computed by using KASH [DF97]. The set of fundamental units of M formed a subset of the set of fundamental units of K. Hence we had $r = 12$ relative units.

Baker's method gave $A < 10^{104}$ for the exponents of the unit equation (cf. (2.5), (3.29)). The reduction algorithm (Lemma 2.2.2) was used with 13 terms in the linear form, according to Table 9.4. In the table we use the notation as in Example 1.

The final bound $A < 86$ implied the bound $S_0 = 10^{2405}$ in (2.9).

Table 9.3:

x_0	x_1	x_2	x_3	x_4	y_0	y_1	y_2	y_3	y_4
15	−3	−27	1	5	−54	15	96	−8	−20
−1	3	−1	−4	2	0	−1	6	−3	0
−262	77	471	−36	−97	−219	65	394	−30	−81
11	−5	−21	2	4	8	−2	−14	1	3
3	−3	−5	1	1	−7	2	14	−1	−3
7	−13	−1	3	0	3	−10	4	2	−1
−6	0	0	0	0	−5	0	0	0	0
0	−1	0	0	0	−2	1	4	0	−1
2	0	0	0	0	−7	0	0	0	0
−2	5	−4	−1	1	1	6	−5	−1	1
−11	0	24	−1	−5	3	1	−5	0	1
4	4	−1	−1	0	3	4	−1	−1	0
−1	2	4	−1	−1	3	−6	−13	2	3
−5	2	9	−1	−2	−3	−1	5	0	−1
1	−3	−4	1	1	3	−2	−9	1	2
3	−3	−9	1	2	1	0	0	0	0
−2	5	−4	−1	1	4	−3	−5	1	1
0	1	1	0	0	0	−1	0	0	0
−3	0	0	0	0	1	0	0	0	0
0	−3	−4	1	1	3	−3	−9	1	2
0	−1	0	0	0	−1	3	4	−1	−1
−2	3	4	−1	−1	2	0	−5	0	1
−1	0	0	0	0	−1	0	0	0	0
0	0	0	0	0	1	0	0	0	0
1	0	0	0	0	0	0	0	0	0

In the enumeration procedure we had 18 ellipsoids in 12 variables. The enumeration of the integer points of the ellipsoids were performed in several steps, according to Table 9.5. In the table we use the notation of Example 1.

Table 9.4:

Step	A_0	H	A_1	Digits	CPUtime
1.	10^{104}	10^{900}	1246	1500	290 min
2.	1246	10^{80}	121	200	19 min
3.	121	10^{60}	91	150	14 min
4.	91	10^{57}	86	150	13 min

Table 9.5:

Step	S	s	Digits	tuples	CPUtime
1.	10^{2405}	10^{50}	200	0	15 min
2.	10^{50}	10^{20}	100	4	6 min
3.	10^{20}	10^{15}	80	8	4 min
4.	10^{15}	10^{12}	80	396	4 min
5.	10^{12}	10^{10}	80	3419	6 min
6.	10^{10}	10^9	80	4574	6 min
7.	10^9	10^8	80	14413	9 min
8.	10^8	10^7	80	39283	18 min
9.	10^7	$5 \cdot 10^6$	80	18093	11 min
10.	$5 \cdot 10^6$	10^6	80	55989	24 min
11.	10^6	$5 \cdot 10^5$	80	33578	16 min
12.	$5 \cdot 10^5$	10^5	80	95078	37 min
13.	10^5	$5 \cdot 10^4$	80	44819	20 min
14.	$5 \cdot 10^4$	10^4	80	113397	43 min
15.	10000	5000	80	38527	20 min
16.	5000	3000	80	27479	14 min
17.	3000	1500	80	27714	14 min
18.	1500	800	80	19034	11 min
19.	800	400	80	14137	9 min
20.	400	200	80	8529	6 min
21.	200	100	80	4447	5 min
22.	100	50	80	1982	3 min
23.	50	25	80	688	2 min
24.	25	10	80	222	2 min
25.	10	3	80	62	1 min
26.	3		80	2	0.5 min

The exponent tuples were tested if there are solutions corresponding to them. The test of the 565869 exponent tuples took about 240 minutes of CPU time.

The element $\alpha \in \mathbb{Z}_K$ generates a relative power integral basis of K over M if and only if it is of the form $\alpha = X_0 + \varepsilon(X_1 \xi + X_2 \xi^2)$ with arbitrary $X_0 \in \mathbb{Z}_M$, an arbitrary unit ε in M and $X_1 = x_0 + x_1 \mu + x_2 \mu^2 + x_3 \mu^3 + x_4 \mu^4 + x_5 \mu^5$, $X_2 = y_0 + y_1 \mu + y_2 \mu^2 + y_3 \mu^3 + y_4 \mu^4 + y_5 \mu^5$ whose coordinates are listed in Table 9.6.

Table 9.6:

x_0	x_1	x_2	x_3	x_4	x_5	y_0	y_1	y_2	y_3	y_4	y_5
12	−150	−139	70	41	−15	−17	45	61	−23	−18	6
−16	−13	8	5	−2	0	5	26	13	−15	−3	2
−33	−13	33	1	−9	2	115	49	−116	−5	32	−7
−3	51	54	−32	−19	8	−1	38	39	−24	−14	6
4	−53	−52	31	17	−7	−1	7	3	−5	1	0
−2	23	8	−8	−2	1	0	−12	1	4	−1	0
−5	7	−6	−2	4	−1	−24	11	50	−12	−16	5
−1	17	19	−8	−6	2	4	−66	−69	33	22	−8
4	−71	−66	40	22	−9	0	−61	−58	34	20	−8
1	11	7	−6	−2	1	0	−31	−29	17	10	−4
0	−3	−7	2	3	−1	−1	28	29	−16	−10	4
0	19	19	−12	−7	3	1	−7	−16	5	6	−2
6	0	0	0	0	0	5	0	0	0	0	0
−3	−6	2	3	−1	0	−1	−11	−7	6	2	−1
0	12	13	−7	−5	2	−1	15	5	−8	−1	1
0	−17	−15	9	5	−2	3	−16	−16	9	5	−2
3	−60	−59	34	20	−8	−1	17	15	−9	−5	2
−2	0	0	0	0	0	7	0	0	0	0	0
1	−4	−9	3	3	−1	−6	−8	3	3	−1	0
−13	−12	15	3	−5	1	−11	−7	6	2	−1	0
2	−13	−15	8	5	−2	−2	5	9	−3	−3	1
−2	−2	1	0	0	0	7	8	−3	−3	1	0
−1	−6	−8	3	3	−1	1	0	0	0	0	0
12	9	−7	−2	1	0	−4	−6	2	3	−1	0
1	7	−2	−3	1	0	−1	12	7	−6	−2	1
1	−3	−7	2	3	−1	0	2	7	−2	−3	1
0	−2	10	0	−4	1	−2	29	−12	−11	7	−1
3	0	0	0	0	0	−1	0	0	0	0	0
1	0	0	0	0	0	−1	12	7	−6	−2	1
1	0	0	0	0	0	1	0	0	0	0	0
0	0	0	0	0	0	1	0	0	0	0	0
1	0	0	0	0	0	0	0	0	0	0	0

Table 9.7:

	reduction	enumeration	test	Total
Example 1	192 min	38.3 min	2 min	3.9 hours
Example 2	336 min	306.5 min	240 min	14.7 hours

9.2.3   Computational experiences

The calculation of the basic data was already a hard problem in the totally real fields of degrees 15, resp. 18 we investigated. Nevertheless, KASH managed this computation in a couple of minutes.

Table 9.7 gives a summary of the CPU times.
A considerable amount of CPU time was taken by the reduction procedure. Proceeding from $r = 10$ to $r = 12$ the reduction times are still comparable but the necessary CPU time for enumeration is about 8 times more. (Note that for $r = 10$ we had 15 ellipsoids, for $r = 12$ we had 18 ellipsoids to enumerate, so the main difference in the CPU times is caused by the difference in the number of variables.) Moreover for $r = 12$ considerable CPU time is taken also by testing the possible exponent vectors, which time was negligible for $r = 10$. These experiences show that $r = 12$ is about the limit of the applicability of the ellipsoid method of Section 2.3.

9.3   Relative quartic extensions

9.3.1   Preliminaries

Relative quartic extensions were considered by I. Gaál and M. Pohst [GP00]. In fact we use here a relative analogue of the method for quartic fields given in Section 6.1. Let K be a quartic extension field of the field M of degree m, generated by an algebraic integer ξ over M, that is $K = M(\xi)$.

We represent any $\alpha \in \mathbb{Z}_K$ in the form

$$\alpha = \frac{1}{d}\left(X_0 + X_1\xi + X_2\xi^2 + X_3\xi^3\right) \tag{9.7}$$

with coefficients $X_i \in \mathbb{Z}_M$ $(1 \leq i \leq 4)$ and with a common denominator $d \in \mathbb{Z}$.
For any primitive element $\alpha \in \mathbb{Z}_K$ we have

$$I(\alpha) = \frac{1}{\sqrt{|D_K|}}\left|\prod_{(i_1,j_1)<(i_2,j_2)}\left(\alpha^{(i_1,j_1)} - \alpha^{(i_2,j_2)}\right)\right|, \tag{9.8}$$

where the pairs (i, j) are ordered lexicographically. The representation (9.7) implies

$$\sqrt{|D_K|} = \frac{1}{i_0}\sqrt{|D_{\mathcal{O}}|} = \frac{1}{i_0}D_M^2\left|\prod_{i=1}^{m}\prod_{1\leq j_1<j_2\leq 4}\left(\xi^{(i,j_1)} - \xi^{(i,j_2)}\right)\right|, \tag{9.9}$$

where $\mathcal{O} = \mathbb{Z}_M[\xi]$ and $i_0 = (\mathbb{Z}_K^+ : \mathcal{O}^+)$.

9.3.2   The cubic relative Thue equation

Let $f(x) = x^4 + A_1x^3 + A_2x^2 + A_3x + A_4 \in \mathbb{Z}_M[x]$ be the relative defining polynomial of ξ over M. Set $H = \{(1, 2, 3, 4), (1, 3, 2, 4), (1, 4, 2, 3)\}$. It is easily seen that for $i = 1, \ldots, m$, the numbers $\vartheta^{(i)}_{j_1 j_2 j_3 j_4} = \xi^{(i,j_1)}\xi^{(i,j_2)} + \xi^{(i,j_3)}\xi^{(i,j_4)}$, $(j_1, j_2, j_3, j_4) \in H$ run through the relative conjugates of a cubic element over $M^{(i)}$. Let

$$F^{(i)}(U, V) = \prod_{(j_1,j_2,j_3,j_4)\in H} \left(U - \vartheta^{(i)}_{j_1 j_2 j_3 j_4} V \right) \in \mathbb{Z}^{(i)}_M.$$

Further, set

$$\begin{aligned}
Q_1(z_1, z_2, z_3) &= z_1^2 - z_1 z_2 A_1 + z_2^2 A_2 + z_1 z_3 (A_1^2 - 2A_2) \\
&\quad + z_2 z_3 (A_3 - A_1 A_2) + z_3^2 (-A_1 A_3 + A_2^2 + A_4), \\
Q_2(z_1, z_2, z_3) &= z_2^2 - z_1 z_3 - z_2 z_3 A_1 + z_3^2 A_2,
\end{aligned}$$

with ternary quadratic forms in $\mathbb{Z}_M[z_1, z_2, z_3]$.

Analogously to the method used in Section 6.1 we have

Theorem 9.3.1 *If α of (9.7) satisfies*

$$I_{K/M}(\alpha) = 1,$$

then there is a solution $(U, V) \in \mathbb{Z}_M$ of

$$N_{M/\mathbb{Q}}(F(U, V)) = \pm \frac{d^{6m}}{i_0} \tag{9.10}$$

such that

$$\begin{aligned}
U &= Q_1(X_1, X_2, X_3), \\
V &= Q_2(X_1, X_2, X_3).
\end{aligned} \tag{9.11}$$

Proof. For any $\alpha \in \mathbb{Z}_K$ we have

$$\prod_{i=1}^{m} \prod_{1 \le j_1 < j_2 \le 4} \left| \frac{d\left(\alpha^{(i,j_1)} - \alpha^{(i,j_2)}\right)}{\left(\xi^{(i,j_1)} - \xi^{(i,j_2)}\right)} \right| = \frac{d^{6m} \cdot (\mathbb{Z}^+_K : \mathbb{Z}^+_M[\alpha])}{i_0}. \tag{9.12}$$

As in Section 6.1 we conclude

$$\prod_{1 \le j_1 < j_2 \le 4} \frac{d\left(\alpha^{(i,j_1)} - \alpha^{(i,j_2)}\right)}{\left(\xi^{(i,j_1)} - \xi^{(i,j_2)}\right)} = F^{(i)}(U^{(i)}, V^{(i)}),$$

which in view of $I_{K/M}(\alpha) = 1$ implies the assertion. $\qquad\square$

Note that concerning the two factors of $I(\alpha)$ in (9.2) in addition to (9.12) we also have

$$\frac{1}{D_M^2} \cdot \prod_{1 \le i_1 < i_2 \le m} \prod_{1 \le j_1 \le 4} \prod_{1 \le j_2 \le 4} d \left| \alpha^{(i_1,j_1)} - \alpha^{(i_2,j_2)} \right|$$
$$= d^{8m(m-1)} \cdot (\mathbb{Z}^+_M[\alpha] : \mathbb{Z}^+[\alpha]). \tag{9.13}$$

If the left-hand side of equation (9.10) is irreducible, then it is a cubic relative Thue equation over M which can be solved by the methods of Section 3.3. Otherwise, if the left-hand side of (9.10) is reducible, then it leads to a very simple unit equation, a case which is treated in detail in Section 10.1.5. In both cases we can determine those pairs $(U, V) \in \mathbb{Z}_M^2$ such that all solutions of (9.10) are of the form $(\eta U, \eta V)$ with a unit $\eta \in \mathbb{Z}_M$. Next we are going to determine the $(X_1, X_2, X_3) \in \mathbb{Z}_M^3$ corresponding to a fixed pair (U, V) by (9.11).

9.3.3 Representing the variables as binary quadratic forms

We continue to follow the arguments of Section 6.1 in a relative sense. For a solution (U, V) of equation (9.10) we set

$$Q_0(z_1, z_2, z_3) = U Q_2(z_1, z_2, z_3) - V Q_1(z_1, z_2, z_3).$$

Using the arguments of C.L. Siegel [Si79] (page 264), it is possible to decide if the equation

$$Q_0(Y_1, Y_2, Y_3) = 0 \quad (Y_1, Y_2, Y_3 \in \mathbb{Z}_M) \tag{9.14}$$

has non-trivial solutions and if so, one can determine a solution (cf. also M. Pohst [Po00]). We assume that (Y_1, Y_2, Y_3) is a non-trivial solution of (9.14) with $Y_3 \neq 0$. (The other possible cases are treated similarly.) Then we can represent any $(X_1, X_2, X_3) \in \mathbb{Z}_M^3$ in the form

$$
\begin{aligned}
X_1 &= Y_1 R + P, \\
X_2 &= Y_2 R + Q, \\
X_3 &= Y_3 R,
\end{aligned}
\tag{9.15}
$$

with $P, Q, R \in M$, $R \neq 0$. From $Q_0(X_1, X_2, X_3) = 0$ we obtain

$$R(C_1 P + C_2 Q) = C_3 P^2 + C_4 P Q + C_5 Q^2 \tag{9.16}$$

with explicitly computable integers $C_1, \ldots, C_5 \in \mathbb{Z}_M$. We multiply the equations of (9.15) by $C_1 P + C_2 Q$ and use (9.16) to eliminate R on the right-hand sides to conclude

$$
\begin{aligned}
S \cdot X_1 &= f_1(P, Q) = C_{11} P^2 + C_{12} P Q + C_{13} Q^2, \\
S \cdot X_2 &= f_2(P, Q) = C_{21} P^2 + C_{22} P Q + C_{23} Q^2, \\
S \cdot X_3 &= f_3(P, Q) = C_{31} P^2 + C_{32} P Q + C_{33} Q^2,
\end{aligned}
\tag{9.17}
$$

again with explicitly given $C_{ij} \in \mathbb{Z}_M$ and with $S = C_1 P + C_2 Q \in M$.

The following statement was immediate in the quartic case, in the relative case it is more complicated:

Lemma 9.3.1 *In order to obtain all integer solutions $(X_1, X_2, X_3) \in \mathbb{Z}_M^3$ of the system (9.11), the parameters P, Q in (9.17) can be replaced by algebraic integer parameters and the corresponding S can have only finitely many non-associated values.*

Proof. Multiplying the equations of (9.17) by the square of a common denominator κ of P and Q yields

$$(C_1\tilde{P} + C_2\tilde{Q})\,\kappa \begin{pmatrix} X_1 \\ X_2 \\ X_3 \end{pmatrix} = C \begin{pmatrix} \tilde{P}^2 \\ \tilde{P}\tilde{Q} \\ \tilde{Q}^2 \end{pmatrix}, \tag{9.18}$$

where $\tilde{P}, \tilde{Q}, \kappa \in \mathbb{Z}_M$, and $C = (C_{ij})$.

Set $(\tilde{P}, \tilde{Q}) = \mathfrak{a}$. Choose an integral ideal $\mathfrak{b}$ of smallest norm in the inverse ideal class of the class of $\mathfrak{a}$. We note that upper bounds for $N(\mathfrak{b})$ are given by R. Zimmert [Zi81]. Since we do not know the class of $\mathfrak{a}$, we need to consider all classes for $\mathfrak{b}$. The ideal $\mathfrak{b}/\mathfrak{a} = (\gamma)$ is a principal ideal. Equation (9.18) multiplied by γ^2 yields

$$s \begin{pmatrix} X_1 \\ X_2 \\ X_3 \end{pmatrix} = C \begin{pmatrix} \hat{P}^2 \\ \hat{P}\hat{Q} \\ \hat{Q}^2 \end{pmatrix}, \tag{9.19}$$

where $\hat{P} = \gamma\tilde{P}, \hat{Q} = \gamma\tilde{Q} \in \mathbb{Z}_M$ and $s \in \mathbb{Z}_M$.

Let $\gcd(X_1, X_2, X_3) = \mathfrak{c} \subseteq \mathbb{Z}_M$. By $Q_1(X_1, X_2, X_3) = U, Q_2(X_1, X_2, X_3) = V$ we have $\mathfrak{c}^2 | \gcd(U, V)$, hence it belongs to a finite set of ideals. Further, $\gcd(sX_1, sX_2, sX_3) = s\,\mathfrak{c} \subseteq \mathbb{Z}_M$, i.e., $s \in \mathfrak{c}^{-1}$.

Set $c = \det(C)$, then $C^{-1} = \dfrac{1}{c}\tilde{C}$, where the 3 by 3 matrix $\tilde{C}$ has entries from $\mathbb{Z}_M$. By equation (9.19) we have

$$s\,\tilde{C} \begin{pmatrix} X_1 \\ X_2 \\ X_3 \end{pmatrix} = c \begin{pmatrix} \hat{P}^2 \\ \hat{P}\hat{Q} \\ \hat{Q}^2 \end{pmatrix}.$$

Determine $s_\mathfrak{b} \in \mathfrak{b}$, resp. $s_\mathfrak{c} \in \mathfrak{c}$ of minimal (absolute) norm. Then $s\,s_\mathfrak{c} \in \mathbb{Z}_M$ and with suitable $g \in \mathbb{Z}_M$

$$s\,s_\mathfrak{c}\,g = c\,s_\mathfrak{c}\,s_\mathfrak{b}^2,$$

also $(s\,s_\mathfrak{c}) | (c\,s_\mathfrak{c}\,s_\mathfrak{b}^2)$. Since the class number of M is finite, (up to unit factors) there are only finitely many possibilities for $s_\mathfrak{c}, s_\mathfrak{b}$, hence also for s. The assertion follows from (9.19) with $S = s$. $\square$

9.3.4 The quartic relative Thue equations

Substituting the representation (9.17) into (9.11) we conclude

$$F_1(P, Q) = Q_1(f_1(P, Q), f_2(P, Q), f_3(P, Q)) = S^2 U, \tag{9.20}$$
$$F_2(P, Q) = Q_2(f_1(P, Q), f_2(P, Q), f_3(P, Q)) = S^2 V. \tag{9.21}$$

By the same formal arguments as in the quartic case (cf. Section 6.1.3) it can be shown that at least one of the equations (9.20), (9.21) is a quartic relative Thue equation over M, where $S^2U, S^2V \in \mathbb{Z}_M$ by the proof of Lemma 9.3.1. (Also, this equation factorizes in the original field K.) That equation can be solved again by the methods of Section 3.3, i.e., we can determine $(P, Q) \in \mathbb{Z}_M$ up to unit factors in M, hence by (9.17) we can calculate $X_1, X_2, X_3 \in \mathbb{Z}_M$ up to a unit factor of M, as well. Thus we can compute those triples (X_{10}, X_{20}, X_{30}) such that the coefficients of any solution α (cf. (9.7)) of $I_{K/M}(\alpha) = 1$ satisfy

$$ X_i = \eta X_{i0} \quad (1 \le i \le 3) $$

with a unit $\eta \in M$. This means that all elements having relative index $I_{K/M}(\alpha) = 1$ are of the form

$$ \alpha = \frac{1}{d}\left(X_0 + \eta(X_{10}\xi + X_{20}\xi^2 + X_{30}\xi^3)\right), $$

where $(X_{10}, X_{20}, X_{30}) \in \mathbb{Z}_M^3$ belongs to a finite set, $X_0 \in \mathbb{Z}_M$ and the unit $\eta \in M$ are arbitrary.

An example of the method applied to a quartic extension of a cubic field is given below. Another application to a quartic extension of a quadratic field is given in Section 10.1.5.

9.3.5 An example for computing relative power integral bases in a field of degree 12 with a cubic subfield

Consider the totally real cubic field M generated by a root ρ of $g(x) = x^3 + 4x^2 - 2x - 1$ having integral basis $\{1, \rho, \rho^2\}$ and discriminant $D_M = 469$. Also, let L be the totally real quartic field generated by a root ξ of $h(x) = x^4 - 4x^2 + x + 1$ having integral basis $\{1, \xi, \xi^2, \xi^3\}$ and discriminant $D_L = 1957$. Let $K = LM = M(\xi)$ be their composite field of degree 12 with the cubic subfield M. The discriminants D_M, D_L being coprime, any element of $\mathbb{Z}_K$ can be written in the form

$$ \alpha = X_0 + X_1\xi + X_2\xi^2 + X_3\xi^3 $$

with $X_0, X_1, X_2, X_3 \in \mathbb{Z}_M$, that is we have $d = 1$ and $i_0 = (\mathbb{Z}_K^+ : \mathcal{O}^+) = 1$. Note that the relative defining polynomial of ξ over M is just $h(x)$. Also, the field K being the composite of L and M makes some of our equations look simpler (e.g., relative Thue equations have coefficients in $\mathbb{Z}$) but these equations are equally hard to solve.

Equation (9.10) has the form

$$ U^3 + 4U^2V - 4UV^2 - 17V^3 = \varepsilon \tag{9.22} $$

with a unit ε in M. This is a cubic relative Thue equation over M, the solutions (U, V) of which can be determined (up to unit factors of M) by the methods of Section 3.3. In the following table we list its solutions together with the corresponding solutions Y_1, Y_2, Y_3 of (9.14). The first column refers to the number of the solution.

	U	V	Y_1	Y_2	Y_3
1.	-2	-1	-5	0	1
2.	1	0	0	2	1
3.	-4	1	0	-1	1
4.	-2	1	-3	0	1

Note that we have unique factorization in M. For each of the above solutions the determinant of the elements C_{ij} in (9.17) is a unit in M, so the element S is also a unit in M. For each solution we list equations (9.17), the quartic equations (9.20) or (9.21) in P, Q, and we display their solutions and the corresponding X_{10}, X_{20}, X_{30}.

Solution 1. In this case we have

$$\begin{aligned} f_1(P, Q) &= 5P^2 + PQ - 30Q^2, \\ f_2(P, Q) &= Q^2, \\ f_3(P, Q) &= -P^2 + 6Q^2, \end{aligned}$$

and

$$F_2(P, Q) = P^4 + P^3 Q - 12 P^2 Q^2 - 6 P Q^3 + 37 Q^4 = \varepsilon,$$
$$(P, Q, X_{10}, X_{20}, X_{30}) = (7, 3, -4, 9, 5), (3, -1, 12, 1, -3),$$
$$(-2, -1, -8, 1, 2), (1, 0, 5, 0, -1).$$

Solution 2. In this case we have

$$\begin{aligned} f_1(P, Q) &= -P^2 + 4PQ, \\ f_2(P, Q) &= -PQ + 2Q^2, \\ f_3(P, Q) &= -Q^2, \end{aligned}$$

and

$$F_1(P, Q) = P^4 - 8P^3 Q + 20 P^2 Q^2 - 15 P Q^3 - Q^4 = \varepsilon,$$
$$(P, Q, X_{10}, X_{20}, X_{30}) =$$
$$(5 - 12\rho - 3\rho^2, 2 - 4\rho - \rho^2, 27 - 29\rho - 8\rho^2, -6 + 3\rho + \rho^2, -8 + 7\rho + 2\rho^2),$$
$$(0, 1, 0, 2, -1), (1, 0, -1, 0, 0), (-2, -1, 4, 0, -1), (-3, -1, 3, -1, -1).$$

Solution 3. In this case we have

$$\begin{aligned} f_1(P, Q) &= -4P^2 - PQ, \\ f_2(P, Q) &= -P^2 - 4PQ - Q^2, \\ f_3(P, Q) &= P^2, \end{aligned}$$

and

$$F_2(P, Q) = P^4 + 9P^3 Q + 18 P^2 Q^2 + 8 P Q^3 + Q^4 = \varepsilon,$$
$$(P, Q, X_{10}, X_{20}, X_{30}) = (-4, 17, 4, -33, 16), (-1, 3, -1, 2, 1),$$
$$(-2, 1, -14, 3, 4), (1, 0, -4, -1, 1), (0, 1, 0, -1, 0).$$

Solution 4. In this case we have

$$
\begin{aligned}
f_1(P, Q) &= -3P^2 - PQ + 6Q^2, \\
f_2(P, Q) &= -Q^2, \\
f_3(P, Q) &= P^2 - 2Q^2,
\end{aligned}
$$

and

$F_2(P, Q) = P^4 - P^3Q - 4P^2Q^2 + 2PQ^3 + 3Q^4 = \varepsilon,$

$(P, Q, X_{10}, X_{20}, X_{30}) = (1 - 4\rho - \rho^2, 1, -10 + \rho + \rho^2, -1, 3 + \rho),$

$(4 - 4\rho - \rho^2, 3 - 4\rho - \rho^2, 2 - 2\rho - \rho^2, -13 + 15\rho + 4\rho^2, -6 + 7\rho + 2\rho^2),$

$(-2, -1, -8, -1, 2), (-1, 1, 4, -1, -1), (1, 1, 2, -1, -1), (1, 0, -3, 0, 1).$

Summarizing, any generator of a relative power integral basis of K over M is of the form

$$
\alpha = X_0 + \varepsilon(X_{10}\xi + X_{20}\xi^2 + X_{30}\xi^3),
$$

where (X_{10}, X_{20}, X_{30}) is one of the solutions listed above, $X_0 \in \mathbb{Z}_M$ and the unit $\varepsilon \in M$ is arbitrary.

10
Some Higher Degree Fields

The resolution of index form equations becomes very difficult for higher degree fields. The method for general quintic fields is already time consuming; for sextic fields a general algorithm does not seem to be feasible, we developed methods for determining power integral bases only in sextic fields having subfields. The case of number fields of degree seven seems to be complicated, since these fields can not have subfields. Special number fields of degree seven (e.g., cyclic fields) can be considered by the methods we used so far.

In this chapter we consider algorithms for the resolution of index form equations in octic and nonic number fields having subfields. Namely, we give an algorithm for solving octic fields with a quadratic subfield using the results of Section 9.3 on relative quartic extensions. Further, we consider nonic fields which are composites of cubic subfields using the results of Section 4.4.1. These algorithms are delicate in the sense that they demonstrate how we can combine several methods to perform the computations. Also, in both cases it took a quite long period of time to find the right ingredients of the method.

At the end of the chapter we also give applications of our previous results to some higher degree fields, even to infinite parametric families of fields.

10.1 Octic fields with a quadratic subfield

10.1.1 Preliminaries

In this section we give an algorithm for determining power integral bases in octic fields with a quadratic subfield, cf. I. Gaál and M. Pohst [GP00]. We utilize the method of Section 9.3 concerning power integral bases in quartic relative extensions. Let K be an octic number field with a quadratic subfield M and assume $K = M(\xi)$ with an algebraic integer ξ. We consider in detail the more interesting case when M is a real quadratic field. Denote by μ the fundamental unit of M with the property $\mu > 1$ and let $\{1, \omega\}$ be an integral basis of M. The conjugate of any $\gamma \in M$ will be denoted by $\overline{\gamma}$. The conjugates of ξ are $\xi^{(i,j)}$, $(1 \le i \le 2, 1 \le j \le 4)$, where $\xi^{(1,j)}$ $(1 \le j \le 4)$ and $\xi^{(2,j)}$ $(1 \le j \le 4)$ are the roots of the relative defining polynomial of ξ over M and of the conjugate of the relative defining polynomial over $\overline{M}$, respectively. The conjugate of any $\gamma \in K$ corresponding to $\xi^{(i,j)}$ is $\gamma^{(i,j)}$ $(1 \le i \le 2, 1 \le j \le 4)$.

We represent the integers $\alpha \in K$ in the form

$$\alpha = \frac{1}{d}(X_0 + X_1 \xi + X_2 \xi^2 + X_3 \xi^3), \tag{10.1}$$

where $d \in \mathbb{Z}$ is a common denominator. Our purpose is now to determine all generators of power integral bases of K, that is elements $\alpha \in \mathbb{Z}_K$ having (absolute) index 1. Obviously,

$$I(\alpha) = (\mathbb{Z}_K^+ : \mathbb{Z}^+[\alpha]) = (\mathbb{Z}_K^+ : \mathbb{Z}_M^+[\alpha]) \cdot (\mathbb{Z}_M^+[\alpha] : \mathbb{Z}^+[\alpha]),$$

hence $I(\alpha) = 1$ implies that both the relative index of α is 1, that is

$$I_{K/M}(\alpha) = (\mathbb{Z}_K^+ : \mathbb{Z}_M^+[\alpha]) = 1 \tag{10.2}$$

and

$$I_M(\alpha) = (\mathbb{Z}_M^+[\alpha] : \mathbb{Z}^+[\alpha]) = 1. \tag{10.3}$$

The elements of K having relative index 1 over M, that is satisfying (10.2), can be determined by using the method of Section 9.3. Applying the arguments of Section 9.3 we can determine those triples $(X_{10}, X_{20}, X_{30}) \in \mathbb{Z}_M^3$ such that the coefficients (X_1, X_2, X_3) of a solution α of $I_{K/M}(\alpha) = 1$ represented in the form (10.1) satisfy

$$X_i = x_i + \omega y_i = \pm \mu^n X_{i0} \quad (1 \le i \le 3), \tag{10.4}$$

where $x_i, y_i \in \mathbb{Z}$ are the coefficients of X_i with respect to the integral basis $\{1, \omega\}$ and $n \in \mathbb{Z}$ is an unknown exponent. Note that to this aim we need to solve a cubic and some corresponding quartic Thue equations over M. The following must be performed for all such triples (X_{10}, X_{20}, X_{30}). We also set $X_0 = x_0 + \omega y_0$ with $x_0, y_0 \in \mathbb{Z}$. Taking conjugates in the above equation we get

$$x_i = \pm \frac{\mu^n X_{i0} \overline{\omega} - (\overline{\mu})^n \overline{X}_{i0} \omega}{\overline{\omega} - \omega},$$

$$y_i = \pm \frac{\mu^n X_{i0} - (\overline{\mu})^n \overline{X}_{i0}}{\omega - \overline{\omega}}, \tag{10.5}$$

for $1 \le i \le 3$ (with the same signs in both equations).

Equations (10.2), (10.3) imply (cf. (9.12) and (9.13))

$$\left| \prod_{i=1}^{2} \prod_{1 \le j_1 < j_2 \le 4} \frac{d\left(\alpha^{(i,j_1)} - \alpha^{(i,j_2)}\right)}{\left(\xi^{(i,j_1)} - \xi^{(i,j_2)}\right)} \right| = \frac{d^{12}}{i_0}, \tag{10.6}$$

and

$$\frac{1}{D_M^2} \cdot \left| \prod_{1 \le j_1 \le 4} \prod_{1 \le j_2 \le 4} d\left(\alpha^{(1,j_1)} - \alpha^{(2,j_2)}\right) \right| = d^{16}, \tag{10.7}$$

respectively, where $i_0 = (\mathbb{Z}_K^+ : \mathbb{Z}_M^+[\xi])$ and in this special case $D_M = (\omega - \overline{\omega})^2$. We note that the product on the left-hand side of (10.7) is a rational integer.

10.1.2 The unit equation

The factors in (10.6) form a complete norm in a field N of degree at most 12. We denote by $\eta_1, \ldots, \eta_s$ a set of fundamental units in N. Then by (10.6) for $1 \le i \le 2$, $1 \le j_1 < j_2 \le 4$ we have

$$\frac{d\left(\alpha^{(i,j_1)} - \alpha^{(i,j_2)}\right)}{\left(\xi^{(i,j_1)} - \xi^{(i,j_2)}\right)} = \rho^{(i,j_1,j_2)} \cdot \gamma^{(i,j_1,j_2)} \cdot \prod_{j=1}^{s} \left(\eta_j^{(i,j_1,j_2)}\right)^{b_j} \tag{10.8}$$

with exponents $b_1, \ldots, b_s \in \mathbb{Z}$, a root of unity $\rho \in N$, and an integer $\gamma \in \mathbb{Z}_N$ of norm $\pm d^{12}/i_0$.

Set $L = M(\alpha^{(1,1)} - \alpha^{(2,1)})$. The field L is of degree $k \in \{4, 8, 16, 32\}$ over M and the product on the left-hand side of (10.7) is a $32/k$-th power of a relative norm of an element ν in $\mathbb{Z}_L$, which is equal to $\pm D_M^2$. This integer can be chosen so that for each conjugate $\nu^{(j_1,j_2)}$ ($1 \le j_1, j_2 \le 4$) the element $\nu^{(j_1,j_2)}$ divides $d(\alpha^{(1,j_1)} - \alpha^{(2,j_2)})$. Consequently, there exists an integer $\beta \in \mathbb{Z}_L$ of norm $\pm d^{16}$ and a root of unity $\zeta \in L$ such that

$$\frac{d(\alpha^{(1,j_1)} - \alpha^{(2,j_2)})}{\nu^{(j_1,j_2)}} = \zeta^{(j_1,j_2)} \cdot \beta^{(j_1,j_2)} \cdot \prod_{j=1}^{r} \left(\varepsilon_j^{(j_1,j_2)}\right)^{a_j}, \tag{10.9}$$

where $\varepsilon_1 \ldots \varepsilon_r$ are fundamental units in L and $a_1, \ldots, a_r \in \mathbb{Z}$.

The following computations must be performed for each tuple $\zeta, \beta, \rho, \gamma$ with the above properties.

We fix pairwise different indices $1 \le j_1, j_2, j_3 \le 4$ and consider Siegel's identity

$$\left(\alpha^{(1,j_1)} - \alpha^{(2,j_2)}\right) + \left(\alpha^{(2,j_2)} - \alpha^{(1,j_3)}\right) + \left(\alpha^{(1,j_3)} - \alpha^{(1,j_1)}\right) = 0,$$

that is

$$\frac{\left(\alpha^{(1,j_1)} - \alpha^{(2,j_2)}\right)}{\left(\alpha^{(1,j_3)} - \alpha^{(2,j_2)}\right)} + \frac{\left(\alpha^{(1,j_3)} - \alpha^{(1,j_1)}\right)}{\left(\alpha^{(1,j_3)} - \alpha^{(2,j_2)}\right)} = 1.$$

We set

$$\phi_{j_1,j_2,j_3} = \frac{\nu^{(j_1,j_2)}\zeta^{(j_1,j_2)}\beta^{(j_1,j_2)}}{\nu^{(j_3,j_2)}\zeta^{(j_3,j_2)}\beta^{(j_3,j_2)}},$$

$$\psi_{j_1,j_2,j_3} = \frac{\left(\xi^{(1,j_3)} - \xi^{(1,j_1)}\right)\rho^{(1,j_3,j_1)}\gamma^{(1,j_3,j_1)}}{\nu^{(j_3,j_2)}\zeta^{(j_3,j_2)}\beta^{(j_3,j_2)}}.$$

By (10.9), (10.8) this equation implies

$$\phi_{j_1,j_2,j_3}\prod_{j=1}^{r}\left(\frac{\varepsilon_j^{(j_1,j_2)}}{\varepsilon_j^{(j_3,j_2)}}\right)^{a_j} + \psi_{j_1,j_2,j_3}\prod_{j=1}^{s}\left(\eta_j^{(1,j_3,j_1)}\right)^{b_j}\prod_{j=1}^{r}\left(\varepsilon_j^{(j_3,j_2)}\right)^{-a_j} = 1.$$

(10.10)

We set $A = \max(|a_1|,\ldots,|a_r|)$ and $B = \max(|a_1|,\ldots,|a_r|,|b_1|,\ldots,|b_s|)$. Obviously, we have $A \le B$. Using standard arguments, and applying Theorem 2.1.1 of A. Baker and G. Wüstholz we obtain

$$\exp(-C \cdot \log A) <$$

$$\left|\log\left(\phi_{j_1,j_2,j_3}\right) + a_1\log\left(\frac{\varepsilon_1^{(j_1,j_2)}}{\varepsilon_1^{(j_3,j_2)}}\right) + \cdots + a_r\log\left(\frac{\varepsilon_r^{(j_1,j_2)}}{\varepsilon_r^{(j_3,j_2)}}\right) + a_0\log(-1)\right|$$

$$< \exp(c_1 - c_2 B) < \exp(c_1 - c_2 A),$$

(10.11)

where log denotes the principal value, C, c_1, c_2 are explicitly given positive constants (C large), and $a_0 \in \mathbb{Z}$ with $|a_0| \le |a_1| + \cdots + |a_r|$. (Note that a_0 can be omitted in the totally real case.) Again it is straightforward to apply a reduction algorithm (cf. Lemma 2.2.2) to the above inequality to get a reduced upper bound for A. We stress that we do not enumerate the possible values of the exponents here, we only need the reduced bound. Note that in fact solving the above unit equation would yield the application of the direct method we mentioned in Section 4.1. The reduction method based on LLL algorithm works fast even in the present situation with a large unit rank, but the enumeration of the exponents under the reduced bound could be troublesome, especially in the totally real case. We rather find another way deriving simpler equations that can be solved easier.

10.1.3 The inhomogeneous Thue equation

Using the representations (10.1) and (10.4) of α, x_i, y_i ($1 \le i \le 3$), equation (10.7) can be written in the form

$$\prod_{1 \le j_1 \le 4}\prod_{1 \le j_2 \le 4}\left(\tau_{1,j_1,j_2}y_0 + \tau_{2,j_1,j_2}\mu^n + \tau_{3,j_1,j_2}(\overline{\mu})^n\right) = \pm d^{16},$$

(10.12)

where τ_{i,j_1,j_2} are algebraic integers in N. Here the only unknowns are n and y_0.

Because of $\mu > 1$, the first and second terms in the factors of equation (10.12) can be large, but the third term involving $(\overline{\mu})^n$ is small in absolute value for $n > 0$.

(The case $n < 0$ is analogous.) Hence, that equation has the same structure as an inhomogeneous Thue equation and can be solved similarly, cf. Section 3.2. However, we do not solve (10.12) for the above triples (X_{10}, X_{20}, X_{30}) separately. Using the reduced upper bound for A from Section 10.1.2 we derive upper bounds for $|n|$ in each case. Namely, according to (10.9) we have

$$\tau_{1,j_1,j_2} y_0 + \tau_{2,j_1,j_2} \mu^n + \tau_{3,j_1,j_2} (\overline{\mu})^n = \frac{d(\alpha^{(1,j_1)} - \alpha^{(2,j_2)})}{\nu^{(j_1,j_2)}}$$

$$= \zeta^{(j_1,j_2)} \cdot \beta^{(j_1,j_2)} \cdot \left(\varepsilon_1^{(j_1,j_2)}\right)^{a_1} \cdots \left(\varepsilon_r^{(j_1,j_2)}\right)^{a_r}. \tag{10.13}$$

Eliminating y_0 from two of these equations and using the reduced upper bound for $A = \max(|a_1|, \ldots, |a_r|)$ we get an upper bound for $|n|$.

Then we only have to enumerate small values of n and test if there are corresponding solutions y_0 of (10.12). For fixed n (10.12) is a polynomial equation in y_0.

10.1.4   Sieving

This upper bound for $|n|$ is usually about 1000–6000, so it is not yet possible to test all potential values of n below this bound directly, especially since μ^n or $(\overline{\mu})^n$ and, accordingly, the coefficients of the polynomial in y_0 which we get from (10.12), can become extremely large.

Hence, we use a *sieve method*. We construct the defining polynomial of a generating element κ of the field L (cf. second paragraph of section 10.1.2). Then we compute primes p such that this minimal polynomial splits completely mod p and p is coprime to the discriminant of L. Consequently, there is a prime ideal $\wp$ in $\mathbb{Z}_L$ lying above p such that for each conjugate of κ we have

$$\kappa^{(j_1,j_2)} \equiv k_{j_1,j_2} \pmod{\wp} \quad (1 \le j_1, j_2 \le 4)$$

for suitable integers k_{j_1,j_2}. By embedding the other elements of L we obtain that there exist integers $t_{i,j_1,j_2}, m, \overline{m}$ such that

$$\tau_{i,j_1,j_2} \equiv t_{i,j_1,j_2} \pmod{\wp} \quad (1 \le i \le 3, \ 1 \le j_1, j_2 \le 4),$$

$$\mu \equiv m \pmod{\wp},$$

$$\overline{\mu} \equiv \overline{m} \pmod{\wp}.$$

Hence, equation (10.12) implies

$$\prod_{1 \le j_1 \le 4} \prod_{1 \le j_2 \le 4} \left(t_{1,j_1,j_2} y_0 + t_{2,j_1,j_2} m^n + t_{3,j_1,j_2} (\overline{m})^n\right) \equiv \pm d^{16} \pmod{p}.$$

$$\tag{10.14}$$

For each n under consideration we can easily decide if that equation has solutions in $y_0 \mod p$. For those exponents for which (10.14) has solutions we use further primes for testing. Extensive computations indicate that after about 10 tests only very few and small exponents remain for which we can solve equation (10.12) in y_0 directly.

10.1.5   An example for computing power integral bases in an octic field with a quadratic subfield

This example illustrates the above method as well as the method of Section 9.3 for determining power integral bases in relative quartic extensions. The relative Thue equations were solved by the algorithm of Section 3.3.

Consider the field K generated by a root ξ of the polynomial

$$f(x) = x^8 - x^7 + x^6 + 2x^5 - 2x^4 + 2x^2 - x - 1.$$

This field has signature (2,6) and discriminant

$$D_K = -4461875 = -5^4 \cdot 11^2 \cdot 59.$$

In the field K both $\{1, \xi, \xi^2, \xi^3, \omega, \omega\xi, \omega\xi^2, \omega\xi^3\}$ and $\{1, \xi, \xi^2, \xi^3, \xi^4, \xi^5, \xi^6, \xi^7\}$ are integral bases, so the denominator d in (10.1) is equal to 1. The field K has $M = \mathbb{Q}(\sqrt{5})$ as a subfield. Set $\omega = \mu = (1 + \sqrt{5})/2$. The relative defining polynomial of ξ over M is $f_M(x) = x^4 + (-1 + \omega)x^3 + x^2 + (1 + \omega)x + \omega$. Equation (9.10) has the form

$$(U + \omega V) \cdot (U^2 + (-1 - \omega)UV + (1 - \omega)V^2) = \varepsilon, \tag{10.15}$$

where ε is a unit in M. The quadratic factor splits in a totally complex quartic field $F = \mathbb{Q}(\gamma)$ containing μ and, say, η as fundamental units. Thus equation (10.15) implies

$$
\begin{aligned}
U + \omega V &= \pm\mu^a, \\
U + \gamma V &= \pm\mu^b\eta^c, \\
U + \overline{\gamma} V &= \pm(\overline{\mu})^b (\overline{\eta})^c,
\end{aligned}
$$

where $\bar{\ }$ denotes the non-trivial automorphism of M and a, b, c are integer exponents. We obtain the unit equation

$$\pm(\omega - \gamma)(\overline{\mu})^b (\overline{\eta})^c \pm (\gamma - \overline{\gamma})\mu^a \pm (\overline{\gamma} - \omega)\mu^b\eta^c = 0.$$

Solving this unit equation we can determine c, whence we get U, V up to a unit factor of M. Together with U, V we also list the corresponding solutions Y_1, Y_2, Y_3 of (9.14) in the following table. The first column refers to the number of the solution.

	U	V	Y_1	Y_2	Y_3
1.	1	0	1	0	1
2.	$-1 - \omega$	-1	0	ω	1
3.	0	$-\omega$	0	1	-1

For each of the previous solutions the determinant of the elements C_{ij} in (9.17) is a unit in M, so the element S is also a unit in M. For each solution we list the equations of (9.17), the quartic equations (9.20) or (9.21) in P, Q, we display their solutions, the corresponding values X_{10}, X_{20}, X_{30}, and we also indicate if there exist exponents n and solutions y_0 of equation (10.12). The element ε denotes a unit in M. In the vectors $(P, Q, X_{10}, X_{20}, X_{30}, n, y_0)$ the $*$ in the last two components means that there are no n, y_0 corresponding to $P, Q, X_{10}, X_{20}, X_{30}$.

Solution 1. In this case we have

$$
\begin{aligned}
f_1(P, Q) &= -P^2 + (1 - \omega)PQ - Q^2, \\
f_2(P, Q) &= -PQ + (1 - \omega)Q^2, \\
f_3(P, Q) &= -Q^2,
\end{aligned}
$$

and

$$F_1(P, Q) = P^4 + (\omega - 1)P^3Q + P^2Q^2 + (\omega + 1)PQ^3 + \omega Q^4 = \varepsilon,$$
$$(P, Q, X_{10}, X_{20}, X_{30}, n, y_0) = (1, -1 + \omega, -5 + 2\omega, 4 - 3\omega, -2 + \omega, *, *),$$
$$(2 + 2\omega, -1 - 3\omega, -12 - 19\omega, 3 + 4\omega, -10 - 15\omega, *, *),$$
$$(-\omega, \omega, -2 - \omega, 1, -1 - \omega, *, *), (0, -1, -1, 1 - \omega, -1, *, *),$$
$$(1, 1, -1 - \omega, -\omega, -1, *, *), (\omega, -1, -1 - \omega, 1, -1, *, *),$$
$$(1, 0, -1, 0, 0, 0, 0), (1, -\omega, -1 - \omega, 0, -1 - \omega, *, *).$$

Solution 2. In this case we have

$$
\begin{aligned}
f_1(P, Q) &= -\omega PQ, \\
f_2(P, Q) &= -\omega P^2 + PQ + Q^2, \\
f_3(P, Q) &= -P^2 + (-1 + \omega)PQ + \omega Q^2,
\end{aligned}
$$

and

$$
\begin{aligned}
&F_2(P, Q) \\
&= (1 + \omega)P^4 - 3\omega P^3Q + (1 - 2\omega)P^2Q^2 + (2 + 2\omega)PQ^3 + (1 + \omega)Q^4 \\
&= (1 + \omega)(P^4 + (3 - 3\omega)P^3Q + (4 - 3\omega)P^2Q^2 + 2PQ^3 + Q^4) = \varepsilon.
\end{aligned}
$$

Note that here the leading coefficient is a unit which can be omitted.

$$(P, Q, X_{10}, X_{20}, X_{30}, n, y_0) = (-2, 2 + \omega, 2 + 6\omega, 1 - \omega, 3 + 6\omega, *, *),$$
$$(-2\omega, -1 - \omega, -4 - 6\omega, -\omega, 1 + 3\omega, *, *), (-1, 1 - \omega, -1, 1 - \omega, 0, 0, 0),$$
$$(1, 0, 0, -\omega, -1, *, *), (-1 + \omega, 1, -1, 1, \omega, *, *),$$
$$(-1, 1 + \omega, 1 + 2\omega, 1 + \omega, 2 + 4\omega, *, *), (-1, 1, \omega, -\omega, 0, -2, 0),$$
$$(0, 1 - \omega, 0, 2 - \omega, -1 + \omega, *, *), (-\omega, 1 + \omega, 2 + 3\omega, -\omega, 1 + 3\omega, *, *).$$

Solution 3. In this case we have

$$
\begin{aligned}
f_1(P, Q) &= \omega P^2, \\
f_2(P, Q) &= -\omega P^2 + (1 + \omega)PQ - \omega Q^2, \\
f_3(P, Q) &= \omega P^2 - PQ + \omega Q^2,
\end{aligned}
$$

and

$$F_2(P, Q)$$
$$= (1+2\omega)P^4 + (-4-6\omega)P^3Q + (6+9\omega)P^2Q^2 + (-4-7\omega)PQ^3$$
$$+ (2+3\omega)Q^4$$
$$= (1+2\omega)(P^4 - 2\omega P^3Q + 3\omega P^2Q^2 + (-2-\omega)PQ^3 + \omega Q^4) = \varepsilon,$$

because $V = -\omega$ is also a unit in M. Note that here the leading coefficient is a unit which can be omitted.

$$(P, Q, X_{10}, X_{20}, X_{30}, n, y_0) = (1, 1, \omega, 1-\omega, -1+2\omega, *, *),$$
$$(-1, -\omega, \omega, -\omega, 1+2\omega, *, *), (-1, 1-\omega, \omega, 1-\omega, \omega, *, *),$$
$$(1, 0, \omega, -\omega, \omega, *, *), (0, 1-\omega, 0, 1-\omega, -1+\omega, *, *).$$

In the general unit equation we had $L = \mathbb{Q}(\xi^{(1,1)} + \xi^{(2,1)}) = \mathbb{Q}(\xi^{(1,1)}\xi^{(2,1)})$. This field is of degree 16, totally complex with unit rank $r = 7$. The defining polynomial of $\xi^{(1,1)}\xi^{(2,1)}$ is

$$x^{16} + x^{15} + x^{14} - 3x^{13} - 3x^{12} - 3x^{11} + 5x^9 + 16x^8 + 24x^7$$
$$+ 22x^6 + 8x^5 - 4x^4 - 6x^3 - 2x^2 + x + 1.$$

Also we put $v^{(j_1,j_2)} = \xi^{(1,j_1)} - \xi^{(2,j_2)}$ and $\zeta = \beta = 1$ in (10.9). Further, by

$$\prod_{i=1}^{2} \prod_{1 \le j_1 < j_2 \le 4} \left(x - \left(\xi^{(i,j_1)} + \xi^{(i,j_2)} \right) \right)$$

$$= (x^4 - x^3 + 2x^2 - 4x + 1)(x^8 - 2x^7 + 3x^5 - 3x^4 + 9x^3 - 9x^2 + 5x - 5)$$

(and similarly for $\xi^{(i,j_1)}\xi^{(i,j_2)}$), we could choose indices (i, j_1, j_2) in (10.8) so that the algebraic number involved lies in a quartic field N, totally complex with unit rank $s = 2$. We took $\rho = \gamma = 1$. Note that for this example Baker's method gave a bound $A \le 10^{49}$ which we reduced in four steps to $A \le 383$. By

$$|n| \le 13.5\, A + 12.87$$

(cf. Section 10.1.3) we got $|n| \le 5183$.

Summarizing, up to equivalence, the following integers generate power integral bases in K:

$$\alpha = \xi, \quad \xi + (-1+\omega)\xi^2, \quad (1-\omega)\xi + (-1+\omega)\xi^2.$$

10.2   Nonic fields with cubic subfields

Having Theorem 4.4.1 on the structure of index form equations in composites of number fields, one is eager to try it for some higher degree fields not covered by the other methods. If in the composite field $K = LM$ the subfield L (resp. M) is a cubic field, then equation (4.9) (resp. equation (4.10)) is a cubic relative Thue equation over the field M (resp. L), that can be solved by the methods of Section 3.3. Why not have both advantages?

Consider a number field K of degree 9 that is the composite of its cubic subfields L, M with coprime discriminants, cf. I. Gaál, [Ga00b]. We develop the algorithm in detail for the case of complex cubic fields, but the main steps of the procedure are applicable in general. The solutions of the cubic relative Thue equations (4.9), (4.10) can be determined up to unit factors of the respective cubic field. By studying a unit equation in the normal closure $\overline{K}$ of K we give a large upper estimate for the unknown exponents involved in the unit factors (we use only Baker's method but do not carry out any computations in $\overline{K}$). Next we construct, exploiting the common variables in the equations (4.9) and (4.10), linear equations involving the unknown unit factors. These linear equations are then used to perform a simple reduction algorithm on the large a priori upper bounds for the exponents in the unknown unit factors, and finally to determine these exponents.

To solve the relative Thue equations (4.9) and (4.10) we need to solve unit equations with only three unknown exponents, whereas if we were using the direct method mentioned in Section 4.1, we had to solve a unit equation resulting from the index form equation over K, involving all 17 fundamental units of the normal closure $\overline{K}$ of K of degree 36 ($\overline{K}$ is totally complex).

The main steps and many ideas of this section are also applicable to algorithms for solving index form equations in other types of composite fields of higher degree.

10.2.1   The relative Thue equations

Let L be a complex cubic field with integral basis $\{1, l_2, l_3\}$ and fundamental unit ε. Let M be a complex cubic field with integral basis $\{1, m_2, m_3\}$ and fundamental unit η. We assume that the discriminants are coprime, $(D_L, D_M) = 1$. (Otherwise we had a common denominator in the representation (10.16) which would result non-unit elements on the right sides of the forcoming equations.) For simplicity we assume also that K has a system of fundamental units of type $\{\varepsilon, \eta, \mu, \nu\}$, since this is almost always the case in numerical examples. (Otherwise there exists a system of independent units of the above type, and the index of this system must be taken into account in the formulas in a straightforward way.)

Denote by $\gamma^{(i)}$, $i = 1, 2, 3$ the conjugates of an element of L or M. We choose the conjugates and the units so that $\varepsilon^{(1)}$ and $\eta^{(1)}$ are real and greater than 1. For any $\gamma \in K$ set $\gamma^{(i,j)}$ for the conjugate corresponding to $\varepsilon^{(i)}, \eta^{(j)}$.

Let $I_L(x, y) \in \mathbb{Z}[x, y]$ be the index form corresponding to the basis $\{l_1 = 1, l_2, l_3\}$ of L, and $I_M(x, y)$ the index form corresponding to the basis $\{m_1 = 1, m_2, m_3\}$ of M. Specializing (4.8) to this situation, we can write any integral element $\alpha \in K$ as

$$\alpha = \sum_{i=1}^{3} \sum_{j=1}^{3} x_{ij} l_i m_j \tag{10.16}$$

with rational integers $x_{i,j}$, $1 \le i, j \le 3$. As an immediate consequence of Theorem 4.4.1 we have:

Lemma 10.2.1 *If α of (10.16) generates a power integral basis in K, then*

$$N_{L/\mathbb{Q}}(I_M(x_{12} + x_{22}l_2 + x_{32}l_3, x_{13} + x_{23}l_2 + x_{33}l_3)) = \pm 1, \quad (10.17)$$
$$N_{M/\mathbb{Q}}(I_L(x_{21} + x_{22}m_2 + x_{23}m_3, x_{31} + x_{32}m_2 + x_{33}m_3)) = \pm 1. \quad (10.18)$$

Since the index forms $I_L(x, y)$, $I_M(x, y)$ are irreducible cubic forms, these equations can be considered as cubic relative Thue equations over cubic fields. At the time of writing the paper [Ga00b], these were the first equations of this type that had been completely solved.

Set

$$X = x_{12} + x_{22}l_2 + x_{32}l_3, \quad Y = x_{13} + x_{23}l_2 + x_{33}l_3. \quad (10.19)$$

Then equation (10.17) is of the form

$$N_{L/\mathbb{Q}}(I_M(X, Y)) = \pm 1 \quad \text{in} \quad X, Y \in \mathbb{Z}_L. \quad (10.20)$$

The solutions of (10.20) can be written as

$$X = \pm \varepsilon^l X_0, \quad Y = \pm \varepsilon^l Y_0 \quad (10.21)$$

with arbitrary $l \in \mathbb{Z}$, where finitely many pairs $(X_0, Y_0) \in \mathbb{Z}_L^2$ are determined by the method of Section 3.3.

Similarly, taking

$$U = x_{21} + x_{22}m_2 + x_{23}m_3, \quad V = x_{31} + x_{32}m_2 + x_{33}m_3, \quad (10.22)$$

equation (10.18) takes the shape

$$N_{M/\mathbb{Q}}(I_L(U, V)) = \pm 1 \quad \text{in} \quad U, V \in \mathbb{Z}_M, \quad (10.23)$$

the solutions of which can be represented in the form

$$U = \pm \eta^k U_0, \quad V = \pm \eta^k V_0, \quad (10.24)$$

with arbitrary $k \in \mathbb{Z}$ and finitely many pairs $(U_0, V_0) \in \mathbb{Z}_M^2$.

10.2.2 The unit equation over the normal closure

We again touch for a moment on the direct method we mentioned in Section 4.1 as an auxiliary tool avoiding calculations in large extension fields of K.

By our assumptions, α with coefficients in (10.16) is a generator of a power integral basis in the composite field K. Taking any $1 \le p_1 < p_2 \le 3, 1 \le q_1 < q_2 \le 3$ by Siegel's identity we have the equation

$$\left(\alpha^{(p_1,q_1)} - \alpha^{(p_1,q_2)} \right) + \left(\alpha^{(p_1,q_2)} - \alpha^{(p_2,q_2)} \right) + \left(\alpha^{(p_2,q_2)} - \alpha^{(p_1,q_1)} \right) = 0.$$

$$(10.25)$$

This equation gives rise to a unit equation in the normal closure $\overline{K}$ of K. Our purpose in considering the above equation is just to give an upper bound for the unknown exponents l and k involved in (10.21) and (10.24). We emphasize that we do not perform any computation in $\overline{K}$ during this phase. The bounds obtained by Baker's method will be reduced by using an efficient algorithm based on independent arguments (cf. Section 10.2.3).

Since we must have $X = \pm\varepsilon^l X_0$, $Y = \pm\varepsilon^l Y_0$, for one of the pairs $(X_0, Y_0) \in \mathbb{Z}_L^2$, using (10.19) we have

$$
\alpha^{(p_1,q_1)} - \alpha^{(p_1,q_2)} = \sum_{i=1}^{3}\sum_{j=1}^{3} x_{ij} l_i^{(p_1)} (m_j^{(q_1)} - m_j^{(q_2)})
$$

$$
= \sum_{j=1}^{3}(m_j^{(q_1)} - m_j^{(q_2)}) \sum_{i=1}^{3} x_{ij} l_i^{(p_1)}
$$

$$
= (m_2^{(q_1)} - m_2^{(q_2)})X^{(p_1)} + (m_3^{(q_1)} - m_3^{(q_2)})Y^{(p_1)}
$$

$$
= \pm\left(\varepsilon^{(p_1)}\right)^l \left((m_2^{(q_1)} - m_2^{(q_2)})X_0^{(p_1)} + (m_3^{(q_1)} - m_3^{(q_2)})Y_0^{(p_1)}\right)
$$

$$
= \pm\left(\varepsilon^{(p_1)}\right)^l \cdot \pi_{p_1,q_1,q_2}
$$

where π_{p_1,q_1,q_2} denotes an algebraic integer from a known finite list.

Similarly, by $U = \pm\eta^k U_0$, $V = \pm\eta^k V_0$ we have (cf. (10.18), (10.22), (10.23), (10.24))

$$
\alpha^{(p_1,q_2)} - \alpha^{(p_2,q_2)} = \sum_{i=1}^{3}\sum_{j=1}^{3} x_{ij}(l_i^{(p_1)} - l_i^{(p_2)})m_j^{(q_2)}
$$

$$
= \sum_{i=1}^{3}(l_i^{(p_1)} - l_i^{(p_2)}) \sum_{j=1}^{3} x_{ij} m_j^{(q_2)}
$$

$$
= (l_2^{(p_1)} - l_2^{(p_2)})U^{(q_2)} + (l_3^{(p_1)} - l_3^{(p_2)})V^{(q_2)}
$$

$$
= \pm\left(\eta^{(q_2)}\right)^k \left((l_2^{(p_1)} - l_2^{(p_2)})U_0^{(q_2)} + (l_3^{(p_1)} - l_3^{(p_2)})V_0^{(q_2)}\right)
$$

$$
= \pm\left(\eta^{(q_2)}\right)^k \cdot \tau_{p_1,p_2,q_2}
$$

with an algebraic integer τ_{p_1,p_2,q_2} from a known finite list.

It follows from the proof of Theorem 4.4.1, that by building $I(\alpha)$ the factors of the square root of $D_K = D_L^3 D_M^3$ in the denominator are absorbed completely in (10.17) and (10.18), and the numbers of type $\kappa = (\alpha^{(p_2,q_2)} - \alpha^{(p_1,q_1)})$ are units themselves. Hence equation (10.25) can be written in the form

$$
\pm\left(\varepsilon^{(p_1)}\right)^l \cdot \pi_{p_1,q_1,q_2} \pm \left(\eta^{(q_2)}\right)^k \cdot \tau_{p_1,p_2,q_2} + \kappa = 0,
$$

that is

$$\pm \left(\frac{\pi_{p_1,q_1,q_2}}{\tau_{p_1,p_2,q_2}}\right)^l \left(\varepsilon^{(p_1)}\right)^l \left(\eta^{(q_2)}\right)^{-k} \pm \left(\frac{1}{\tau_{p_1,p_2,q_2}}\right) \kappa \left(\eta^{(q_2)}\right)^{-k} = 1.$$

This is a unit equation in $\overline{K}$. Applying the effective estimates of Y. Bugeaud and K. Győry [BGy96a] one can easily get an upper bound for the heights of the solutions of this unit equation. For this purpose one only needs to have an upper bound for the heights of the coefficients of the unit equation (this can be easily calculated), and for the degree, unit rank and discriminant of $\overline{K}$. The discriminants of the normal closures $\overline{L}, \overline{M}$ of L, resp. M can also be easily determined. Since in our examples (which involve complex cubic fields L and M) these are fields of degree 6 with coprime discriminants and $\overline{K} = \overline{L}\,\overline{M}$, we have

$$D_{\overline{K}} = \left(D_{\overline{L}}\right)^6 \cdot \left(D_{\overline{M}}\right)^6.$$

Moreover, $\overline{K}$ is a totally complex field of degree 36 with unit rank 17. Using the result of Y. Bugeaud and K. Győry [BGy96a] we obtained an upper bound for the height of $\varepsilon^l \cdot \eta^{-k}$ which allows us to derive an upper bound for $\max(|k|, |l|)$. In our examples this bound was about 10^{247}. Note that we could not explicitly calculate the fundamental units of $\overline{K}$ and used only general upper estimates for their heights (in terms of the degree and discriminant), which made our bounds much weaker.

10.2.3 The common variables

Using (10.19) and (10.21), we can represent $x_{12}, x_{22}, x_{32}, x_{13}, x_{23}, x_{33}$ in terms of the only unknown l. Similarly, using (10.22) and (10.24), we can represent $x_{21}, x_{22}, x_{23}, x_{31}, x_{32}, x_{33}$ in terms of the only unknown k. Hence, for $x_{22}, x_{23}, x_{32}, x_{33}$ we have two different representations, which enables us to relate the unknown exponents k and l.

For any cubic algebraic numbers β, γ, δ (of L or M) let us introduce the notation

$$|\beta, \gamma, \delta| = \begin{vmatrix} \beta^{(1)} & \gamma^{(1)} & \delta^{(1)} \\ \beta^{(2)} & \gamma^{(2)} & \delta^{(2)} \\ \beta^{(3)} & \gamma^{(3)} & \delta^{(3)} \end{vmatrix}.$$

Using the three embeddings of each of the following equations

$$\begin{aligned} X &= & x_{12} + x_{22}l_2 + x_{32}l_3 &= \pm\varepsilon^l X_0, \\ Y &= & x_{13} + x_{23}l_2 + x_{33}l_3 &= \pm\varepsilon^l Y_0, \\ U &= & x_{21} + x_{22}m_2 + x_{23}m_3 &= \pm\eta^k U_0, \\ V &= & x_{31} + x_{32}m_2 + x_{33}m_3 &= \pm\eta^k V_0, \end{aligned} \qquad \begin{aligned} &(10.26) \\ \\ &(10.27) \end{aligned}$$

by Cramer's rule we obtain

$$\pm\frac{|1,\varepsilon^l X_0, l_3|}{|1,l_2,l_3|} = x_{22} = \pm\frac{|1,\eta^k U_0, m_3|}{|1,m_2,m_3|},$$

$$\pm\frac{|1,l_2,\varepsilon^l X_0|}{|1,l_2,l_3|} = x_{32} = \pm\frac{|1,\eta^k V_0, m_3|}{|1,m_2,m_3|},$$

$$\pm\frac{|1,\varepsilon^l Y_0, l_3|}{|1,l_2,l_3|} = x_{23} = \pm\frac{|1,m_2,\eta^k U_0|}{|1,m_2,m_3|},$$

$$\pm\frac{|1,l_2,\varepsilon^l Y_0|}{|1,l_2,l_3|} = x_{33} = \pm\frac{|1,m_2,\eta^k V_0|}{|1,m_2,m_3|}.$$

By expanding the determinants we obtain a system of equations of type

$$\delta_{i1}\left(\varepsilon^{(1)}\right)^l + \delta_{i2}\left(\varepsilon^{(2)}\right)^l + \delta_{i3}\left(\varepsilon^{(3)}\right)^l$$
$$+\delta_{i4}\left(\eta^{(1)}\right)^k + \delta_{i5}\left(\eta^{(2)}\right)^k + \delta_{i6}\left(\eta^{(3)}\right)^k = 0$$

for $i = 1, 2, 3, 4$, with explicitly given algebraic coefficients $\delta_{ij} \in \mathbb{C}$. In almost all cases (remember that there is a list of possibilities for X_0, Y_0, U_0, V_0) this system of equations allows us to eliminate the conjugates of ε^l and to derive an equation of type

$$\delta_1\left(\eta^{(1)}\right)^k + \delta_2\left(\eta^{(2)}\right)^k + \delta_3\left(\eta^{(3)}\right)^k = 0 \qquad (10.28)$$

(with known algebraic coefficients δ_i). After dividing by one of the summands, this equation becomes a very simple unit equation to which the Baker's method (cf. Section 2.1) and reduction (cf. Section 2.2) can be applied to derive a reasonable upper bound for k.

In the remaining cases the original system of four linear equations has only rank 3, but it was always possible to express the conjugates of ε^l as a linear combination of the conjugates of η^k, that is to calculate coefficients δ_{ij} such that

$$\left(\varepsilon^{(i)}\right)^l = \delta_{i1}\left(\eta^{(1)}\right)^k + \delta_{i2}\left(\eta^{(2)}\right)^k + \delta_{i3}\left(\eta^{(3)}\right)^k, \qquad (i = 1, 2, 3). \qquad (10.29)$$

Recall that $\varepsilon^{(1)}$ and $\eta^{(1)}$ are real and greater than 1, while the other conjugates are complex conjugate pairs, with absolute value less than 1.

If $k \geq 0, l \geq 0$ or $k \geq 0, l < 0$, then equation (10.29) allows us to derive easily an upper bound for k.

If $k < 0, l \geq 0$ then

$$\left|\delta_{22}\left(\eta^{(2)}\right)^k + \delta_{23}\left(\eta^{(3)}\right)^k\right| \leq 1 + |\delta_{21}|, \qquad (10.30)$$

where in our applications we always had $\delta_{22} = \overline{\delta_{23}}$ (bar denotes complex conjugation), and we get a similar inequality for $k < 0, l < 0$.

Consider now inequality (10.30). It can be written as

$$\left| \delta \left(\eta^{(2)} \right)^k + \overline{\delta} \left(\overline{\eta^{(2)}} \right)^k \right| \leq 1 + |\delta_{21}| = c_1,$$

with $\delta = \delta_{22}$. We have

$$2|\delta| \left| \eta^{(2)} \right|^k \cdot |\cos(\vartheta_\delta + k\vartheta_{\eta^{(2)}})| \leq c_1, \tag{10.31}$$

where ϑ_δ and $\vartheta_{\eta^{(2)}}$ denote the arguments of the complex numbers appearing as subscripts. The problem is that this cosine function can sometimes become very small (depending on k).

Now if for some integer n we have

$$\left| \vartheta_\delta + k\vartheta_{\eta^{(2)}} - \frac{(2n+1)\pi}{2} \right| > 0.1, \tag{10.32}$$

then

$$2|\delta| \left| \eta^{(2)} \right|^k \leq \frac{c_1}{0.09983},$$

which implies an upper bound for $|k|$ (remember that $k < 0, |\eta^{(2)}| < 1$). If (10.32) is not satisfied, then from the Taylor expansion of the cosine function at $(2n+1)\pi/2$ we obtain

$$\cos(\vartheta_\delta + k\vartheta_{\eta^{(2)}}) = \quad -(-1)^n \cdot \left(\vartheta_\delta + k\vartheta_{\eta^{(2)}} - \frac{(2n+1)\pi}{2} \right)$$
$$-\frac{\cos(\xi)}{2} \left(\vartheta_\delta + k\vartheta_{\eta^{(2)}} - \frac{(2n+1)\pi}{2} \right)^2,$$

where ξ is an intermediate value, hence $|\cos(\xi)| \leq 0.1$ and

$$\left|\cos(\vartheta_\delta + k\vartheta_{\eta^{(2)}})\right| \geq 0.95 \cdot \left| \vartheta_\delta + k\vartheta_{\eta^{(2)}} - \frac{(2n+1)\pi}{2} \right|,$$

which combined with (10.31) yields

$$\left| \vartheta_\delta + k\vartheta_{\eta^{(2)}} - \frac{(2n+1)\pi}{2} \right| \leq c_2 \cdot c_3^{-|k|}, \tag{10.33}$$

where

$$c_2 = \frac{c_1}{1.9\,|\delta|}, \quad c_3 = \frac{1}{|\eta^{(2)}|} > 1.$$

Set

$$\vartheta = \frac{\vartheta_{\eta^{(2)}}}{\pi}, \quad \beta = \frac{2\vartheta_{\delta_2} - \pi}{2\pi},$$

then (10.33) gives

$$\big|\ |k|\,\vartheta + n - \beta\ \big| \leq \frac{c_2}{\pi} \cdot c_3^{-|k|} \leq c_4 \cdot c_3^{-H}, \tag{10.34}$$

where $c_4 = c_2/\pi$ and $H = \max(|k|, |n|)$. This inequality can be used to perform a reduction algorithm for k by using the Davenport lemma (cf. Lemma 2.2.1). The upper bound for $|k|$ was at most 50 in our examples.

For each k with absolute value under this bound we proceed as follows. Using (10.27) we determine $x_{22}, x_{23}, x_{32}, x_{33}$ corresponding to each value of k. Then we use (10.26) to derive an upper bound for l from the values of these common variables (eliminating x_{12} from the first equation or x_{13} from the second equation of (10.26) by subtracting two of its conjugates).

Once we have established an interval for the possible values of l corresponding to k (this interval is usually very short), using (10.26) and (10.27) we calculate the values of the common variables $x_{22}, x_{23}, x_{32}, x_{33}$ corresponding to k and l. If they coincide, we calculate also $x_{12}, x_{13}, x_{21}, x_{31}$, substitute them into (10.16) and test if α does indeed have index 1.

10.2.4   Examples

Using the algorithm described above we made calculations for the following three examples:

$$
\begin{aligned}
&1. \quad f(x) = x^3 - x + 1, & D_L &= -23, \\
&\quad\quad g(x) = x^3 - 2x + 2, & D_M &= -76; \\
&2. \quad f(x) = x^3 + x + 1, & D_L &= -31, \\
&\quad\quad g(x) = x^3 + x^2 + x + 2, & D_M &= -83; \\
&3. \quad f(x) = x^3 + 2x + 1, & D_L &= -59, \\
&\quad\quad g(x) = x^3 + x^2 - 2x - 3, & D_M &= -87;
\end{aligned}
$$

where f and g denote the minimal polynomials of generating elements of L and M, respectively.

In each step of the algorithm we had several solutions, even for the x_{ij} in the last step: for several k there were corresponding values of l representing the same common variables. However, in each of our examples the final test eliminated all candidates and finally there were no elements in the field $K = LM$ having index 1.

A direct search for elements of small index in nonic fields of the considered type seems to make plausible the experience of our examples: such fields seldom have power integral bases.

10.3   Some more fields of higher degree

In this section we give applications of former results to some types of higher degree fields. Some of these applications concern infinite parametric families of fields.

10.3.1   Power integral bases in imaginary quadratic extensions of totally real cyclic fields of prime degree

Let p denote an odd prime. Let L be a totally real cyclic number field of degree p, with integral basis $\{l_1 = 1, l_2, \ldots, l_p\}$, and discriminant D_L. Denote by $I_L(x_2, \ldots, x_p)$ the index form corresponding to the integral basis $\{l_1 = 1, l_2, \ldots, l_p\}$. Also, let $0 < m \in \mathbb{Z}$ be square free with $m \neq 1, 3$ and let $M = \mathbb{Q}(i\sqrt{m})$. An integral basis of M is given by $\{1, \omega\}$ with

$$\omega = \begin{cases} (1 + i\sqrt{m})/2 & \text{if } -m \equiv 1 \ (\mathrm{mod} \ 4), \\ i\sqrt{m} & \text{if } -m \equiv 2, 3 \ (\mathrm{mod} \ 4). \end{cases} \tag{10.35}$$

The discriminant of M is

$$D_M = \begin{cases} -m & \text{if } -m \equiv 1 \ (\mathrm{mod} \ 4), \\ -4m & \text{if } -m \equiv 2, 3 \ (\mathrm{mod} \ 4). \end{cases} \tag{10.36}$$

As above, we assume that $(D_L, D_M) = 1$. Power integral bases in the field $K = LM$ were considered by I. Gaál [Ga98a]. The integers of K can be represented in the form

$$\alpha = x_1 + x_2 l_2 + \cdots + x_p l_p + y_1 \omega + y_2 \omega l_2 + \cdots + y_p \omega l_p \tag{10.37}$$

with $x_j, y_j \in \mathbb{Z}$, $(1 \leq j \leq p)$.

Theorem 10.3.1 *Assume $m \neq 1, 3$ and $(D_L, D_M) = 1$. If the integer α of (10.37) generates a power integral basis in $K = LM$, then*

$$I_L(x_2, \ldots, x_p) = \pm 1, \tag{10.38}$$

$y_1 = \pm 1$ *and* $y_2 = \cdots = y_p = 0$.

In other words, α must be of the form $\alpha = \beta \pm \omega$ with $\beta \in \mathbb{Z}_L$, where β generates a power integral basis in L.

The converse of the assertion is of course not true: elements of the above type do not necessarily generate a power integral basis in K.

Before proving the theorem, we formulate an important consequence of it:

Corollary 10.3.1 *Let $p \geq 5$ and assume as above $m \neq 1, 3$ and $(D_L, D_M) = 1$. If L is not the maximal real subfield of a cyclotomic field, then the composite field $K = LM$ admits no power integral bases.*

Proof of the Corollary. In view of Theorem 1.2.1 of M.N. Gras, the cyclic field L of prime degree $p \geq 5$ can only have power integral bases if L is the maximal real subfield of a cyclotomic field. Hence equation (10.38) is unsolvable in other cases.   □

Now we turn to the proof of Theorem 10.3.1.

Proof of Theorem 10.3.1. Assume that α of (10.37) generates a power integral basis in K. Set $X_j = x_j + \omega y_j$ for $2 \leq j \leq p$. As an application of Theorem 4.4.1 we have

$$N_{M/\mathbb{Q}}(I_L(X_2, \ldots, X_p)) = \pm 1, \qquad (10.39)$$

$$N_{L/\mathbb{Q}}(y_1 + y_2 l_2 + \cdots + y_p l_p) = \pm 1. \qquad (10.40)$$

By our assumption on m, the unit group of M is trivial, hence equation (10.39) implies

$$I_L(X_2, \ldots, X_p) = \pm 1. \qquad (10.41)$$

We shall now show that the unit groups of K and L coincide. Obviously, the unit ranks are equal. By considering those n for which $\varphi(n)$ divides $2p = [K : \mathbb{Q}]$, one can see that if $m \neq 1, 3$ and K is not the cyclotomic field of degree $2p$, where $p_1 = 2p+1$ is prime, then apart from ± 1 there are no other torsion units in K. The assumption $(D_L, D_M) = 1$ excludes that K is the above cyclotomic field, for in that case both D_L and D_M were divisible by $p_1 = 2p+1$. Denote by $\varepsilon_1, \ldots, \varepsilon_{p-1}$ a set of fundamental units in L. It is sufficient to show that for any η of the form

$$\eta = \pm \varepsilon_1^{a_1} \ldots \varepsilon_{p-1}^{a_{p-1}} \qquad (10.42)$$

with $0 \leq a_j \leq 1$ $(1 \leq j \leq p - 1)$, the square root of η is not contained in K. Suppose on the contrary that $\sqrt{\eta} \in K$. Then there exist $\gamma, \delta \in L$ such that

$$\sqrt{\eta} = \gamma + \delta i \sqrt{m},$$

that is

$$\eta = \gamma^2 - m\delta^2 + 2i\gamma\delta\sqrt{m}.$$

By comparing the imaginary parts, it follows that $\gamma\delta = 0$. If $\delta = 0$, then $\sqrt{\eta} = \gamma \in L$ contradicts the fact that $\varepsilon_1, \ldots, \varepsilon_{p-1}$ is a set of fundamental units in L. Assume now that $\gamma = 0$, and let $d \in \mathbb{Z}$ be such that $\delta_0 = d\delta$ is integer in L. Then we get

$$\eta = -m \frac{\delta_0^2}{d^2},$$

hence

$$a = -\frac{d^2}{m} = \frac{\delta_0^2}{\eta}$$

is an integer in $\mathbb{Z}$ because the right-hand side is an integer. By taking the norm it follows that

$$a^p = \pm (N_{L/\mathbb{Q}}(\delta_0))^2,$$

which is impossible for $p > 2$ except for $a = \pm 1$ in which case $\eta = \pm \delta_0^2$ contradicts again the fact that $\varepsilon_1, \ldots, \varepsilon_{p-1}$ are fundamental units in L.

Consider now equation (10.41). As is known, the index form $I_L(X_2, \ldots, X_p)$ can be factorized into linear factors $f_j(X_2, \ldots, X_p)$ $(1 \leq j \leq p(p-1)/2)$ with algebraic integer coefficients. The field L being cyclic, the coefficients of the linear forms are contained in L. If α is a generator of a power integral basis in K, then

$$\prod_{j=1}^{p(p-1)/2} f_j(X_2, \ldots, X_p) = \pm 1.$$

Thus each linear factor is a unit in K. But the unit groups of K and L coincide, hence each linear factor is also a unit in L:

$$f_j(X_2, \ldots, X_p) = \eta_j \qquad (1 \le j \le p(p-1)/2),$$

with some units $\eta_j \in L$. Subtracting the conjugate over M of each linear factor from itself we obtain

$$f_j(y_2, \ldots, y_p) = 0 \qquad (1 \le j \le p(p-1)/2).$$

As it is known, the rank of the system of linear forms f_j, $1 \le j \le p(p-1)/2$ is $p - 1$, hence the above system of equations implies $y_2 = \ldots = y_p = 0$. By this and equation (10.41) we get (10.38). Also, $y_1 = \pm 1$ follows from (10.40). $\qquad\square$

10.3.2 Power integral bases in imaginary quadratic extensions of Lehmer's quintics

Let n be an integer parameter and consider the family of totally real cyclic quintic fields $L = \mathbb{Q}(\vartheta)$ generated by a root of the polynomial (7.12) in Section 7.2. Set $m_0 = n^4 + 5n^3 + 15n^2 + 25n + 25$ and $d = n^3 + 5n^2 + 10n + 7$ as in (7.13), assume that m_0 is square free. By Lemma 7.2.1 an integral basis of L is given by $\{1, \vartheta, \vartheta^2, \vartheta^3, \omega_5\}$ with ω_5 defined there. Recall that $D_L = m_0^4$.

Let $0 < m \in \mathbb{Z}$ be square free with $m \ne 1, 3$ and let $M = \mathbb{Q}(i\sqrt{m})$. Let ω and D_M be the same as in (10.35), (10.36).

Under the above conditions we have (cf. I. Gaál [Ga98a]):

Theorem 10.3.2 *If m_0 is coprime to D_M, then the field $K = \mathbb{Q}(\vartheta, i\sqrt{m})$ contains no power integral bases.*

Proof. Under the above conditions the discriminants of L and M are coprime. An integral basis of K is given by $\{1, \vartheta, \vartheta^2, \vartheta^3, \omega_5, \omega, \omega\vartheta, \omega\vartheta^2, \omega\vartheta^3, \omega\omega_5\}$.

Denote by $I_L(x_2, \ldots, x_5)$ the index form corresponding to the integral basis $\{1, \vartheta, \vartheta^2, \vartheta^3, \omega_5\}$ of L. By Theorem 10.3.1, if

$$\alpha = x_1 + x_2\vartheta + x_3\vartheta^2 + x_4\vartheta^3 + x_5\omega_5 + y_1\omega + y_2\omega\vartheta + y_3\omega\vartheta^2 + y_4\omega\vartheta^3 + y_5\omega\omega_5$$

$(x_j, y_j \in \mathbb{Z})$ generates a power integral basis in K, then

$$I_L(x_2, \ldots, x_5) = \pm 1 \quad \text{in} \quad x_2, x_3, x_4, x_5 \in \mathbb{Z}.$$

By Theorem 7.2.1 this equation is only solvable for $n = -1, -2$. Hence K can have power integral bases only for $n = -1, -2$. Since these define the same field, let us fix $n = -1$.

Again, applying Theorem 10.3.1 in case $n = -1$, if $\alpha \in K$ generates a power integral basis, then

$$\alpha = x_1 + x_2\vartheta + x_3\vartheta^2 + x_4\vartheta^3 + x_5\omega_5 \pm \omega,$$

where $x_1 \in \mathbb{Z}$ is arbitrary and (x_2, x_3, x_4, x_5) is listed at the end of Section 7.2.4. We tested these values of α directly and found no generators of power integral bases.   □

10.3.3 One more composite field

In this Section we apply the results of Section 4.4.2. This is an example to illustrate that Theorem 4.4.2 is easy to use for certain higher degree fields.

Let φ be a root of $f(x) = x^5 - 2x^4 + 7x^2 + 6x + 5$. The quintic field $L = \mathbb{Q}(\varphi)$ has no non-trivial subfields. Let ψ be a root of $g(x) = x^8 + 13x^7 + 55x^6 + 75x^5 + 2x^3 - x^2 - 143x - 525$. The octic field $M = \mathbb{Q}(\psi)$ has no non-trivial subfields, either. We have

$$
\begin{aligned}
f(x) &\equiv (x + 16)^2(x^3 + 16x + 5) \pmod{17}, \\
g(x) &\equiv (x + 5)^2(x^3 + 12x^2 + 2x + 14)(x^3 + 8x^2 + 4x + 7) \pmod{17},
\end{aligned}
$$

hence our Theorem 4.4.2 applies. Consider the order $\mathcal{O}_{fg} = \mathbb{Z}[\varphi, \psi]$ of the field $K = \mathbb{Q}(\varphi, \psi)$ of degree 40. Any $\alpha \in \mathcal{O}_{fg}$ can be represented in the form

$$\alpha = \sum_{i=0}^{4} \sum_{j=0}^{7} x_{ij} \varphi^i \psi^j$$

with $x_{ij} \in \mathbb{Z}$. By Theorem 4.4.2 the indices of all primitive elements of $\mathcal{O}_{fg}$ are divisible by 17, hence $\mathcal{O}_{fg}$ admits no power integral bases.

11

Tables

At the end of the book we provide the reader with tables containing generators of power integral bases. For cubic, quartic and sextic fields, our fast algorithms enables us to list the generators of power integral bases for all number fields with small discriminants. We give the data usually in increasing order of discriminants. These data complete other number field data contained in similar tables. Recall, that in the more complicated fields, where the algorithms require tedious computations, some examples were given in the corresponding sections.

The table of Section 11.1 gives all generators of power integral bases in about 130 cubic fields with small discriminants (both positive and negative).

Section 11.2 is devoted to tables of quartic fields. Here we also give tables that summarize the behaviour of minimal indices of quartic fields as the discriminant grows. We consider cyclic quartic fields in Section 11.2.3. The complete lists of totally real and totally complex biquadratic number fields up to discriminants 10^6 and 10^4 respectively, with all possible generators of power integral bases are given in Sections 11.2.5, 11.2.6, respectively. Further, dozens of quartic fields are given of the remaining signatures and Galois groups.

The five totally real cyclic sextic fields with smallest discriminants are listed in Section 11.3.1 with all generators of power integral bases. The 25 sextic fields with an imaginary quadratic subfield with smallest discriminants (in absolute value) and their generators of power integral bases are given in Section 11.3.2.

11.1   Cubic fields

The table below contains all power integral bases of cubic fields of discriminants $-300 \le D_K \le 3137$. The rows contain the following data:

$$D_K, \quad (a_1, a_2, a_3), \quad (I_0, I_1, I_2, I_3), \quad (x, y), \ldots$$

where D_K is the discriminant of the field K, generated by a root ϑ of the polynomial $f(x) = x^3 + a_1 x^2 + a_2 x + a_3$. In most of these fields $\{1, \omega_2 = \vartheta, \omega_3 = \vartheta^2\}$ is an integral basis; if not then an integral basis is given by $\{1, \omega_2, \omega_3\}$ with $\omega_2 = (p_0 + p_1\vartheta + p_2\vartheta^2)/p$, $\omega_3 = (q_0 + q_1\vartheta + q_2\vartheta^2)/q$ and the table includes the coefficients $\omega_2 = (p_0, p_1, p_2)/p$, $\omega_3 = (q_0, q_1, q_2)/q$. Finally, the solutions $(x, y), \ldots$ of the index form equation

$$I(x, y) = I_0 x^3 + I_1 x^2 y + I_2 x y^2 + I_3 y^3 = \pm 1$$

are displayed. All generators of power integral bases of the field K are of the form

$$\alpha = a \pm (x\omega_2 + y\omega_3),$$

where $a \in \mathbb{Z}$ is arbitrary and (x, y) is a solution of the above index form equation.

11.1.1 Totally real cubic fields

D_K	$f(x)$	$I(x, y)$	solutions
49	$(-1, -2, 1)$	$(1, 2, -1, -1)$	$(-2, 1), (-9, 4), (-1, 1), (-1, -1), (-1, 2)$
			$(-5, 9), (0, 1), (4, 5), (1, 0)$
81	$(0, -3, 1)$	$(1, 0, -3, 1)$	$(-2, 1), (1, 1), (3, 2), (0, 1), (1, 3), (1, 0)$
148	$(-1, -3, 1)$	$(1, 2, -2, -2)$	$(1, 1), (1, -1), (-5, 2), (-31, 45), (1, 0)$
169	$(-1, -4, -1)$	$(1, 2, -3, -5)$	$(-2, 1), (-1, 1), (1, 0)$
229	$(0, -4, 1)$	$(1, 0, -4, 1)$	$(-2, 1), (-2, -1), (508, 273), (0, 1), (1, 4),$
			$(1, 0)$
257	$(-1, -4, 3)$	$(1, 2, -3, -1)$	$(1, 1), (6, 5), (-3, 1), (0, 1), (-2, 7), (1, 0)$
316	$(-1, -3, 2)$	$(1, 2, -3, -2)$	$(-1, 2), (1, 0)$
321	$(-1, -4, 1)$	$(1, 2, -3, -3)$	$(1, -1), (1, 0)$
361	$(-1, -6, 7)$	$(1, 2, -5, 1)$	$(-7, 2), (1, 1), (9, 7), (0, 1), (2, 9), (1, 0)$
404	$(-1, -5, -1)$	$(1, 2, -4, -6)$	$(1, -1), (1, 0)$
469	$(-1, -5, 4)$	$(1, 2, -4, -1)$	$(0, 1), (1, 0)$
473	$(0, -5, 1)$	$(1, 0, -5, 1)$	$(-2, -1), (-7, 3), (0, 1), (1, 5), (1, 0)$
564	$(-1, -5, 3)$	$(1, 2, -4, -2)$	$(3, 2), (-3, 1), (-3, 7), (1, 0)$
568	$(-1, -6, -2)$	$(1, 2, -5, -8)$	$(17, 8), (1, 0)$
621	$(0, -6, 3)$	$(1, 0, -6, 3)$	$(-8, 3), (2, 1), (1, 2), (1, 0)$
697	$(0, -7, 5)$	$(1, 0, -7, 5)$	$(-3, 1), (2, 1), (13, 6), (1, 1), (1, 0)$
733	$(-1, -7, 8)$	$(1, 2, -6, 1)$	$(3, 2), (0, 1), (1, 0)$
756	$(0, -6, 2)$	$(1, 0, -6, 2)$	$(1, 3), (1, 0)$
761	$(-1, -6, -1)$	$(1, 2, -5, -7)$	$(2, 1), (-3, 1), (-1, 1), (1, 0)$
785	$(-1, -6, 5)$	$(1, 2, -5, -1)$	$(0, 1), (1, 0)$
788	$(-1, -7, -3)$	$(1, 2, -6, -10)$	$(-3, 1), (-3, 2), (1, 0)$
837	$(0, -6, 1)$	$(1, 0, -6, 1)$	$(0, 1), (1, 6), (1, 0)$
892	$(-1, -8, 10)$	$(1, 2, -7, 2)$	$(1, 0)$
940	$(0, -7, 4)$	$(1, 0, -7, 4)$	$(1, 0)$
961	$(-1, -10, 8)$	$(2, 5, -1, -2)$	no solutions
	$\omega_2 = (0, 1, 0)/1$	$\omega_3 = (0, 1, 1)/2$	
985	$(-1, -6, 1)$	$(1, 2, -5, -5)$	$(1, -1), (2, 1), (-3, 1), (1, 0)$
993	$(-1, -6, 3)$	$(1, 2, -5, -3)$	$(-1, 2), (1, 0)$
1016	$(-1, -6, 2)$	$(1, 2, -5, -4)$	$(1, 0)$
1076	$(0, -8, 6)$	$(1, 0, -8, 6)$	$(7, 3), (1, 1), (1, 0)$
1101	$(-1, -9, 12)$	$(1, 2, -8, 3)$	$(1, 0)$

D_K	$f(x)$	$I(x,y)$	solutions
1129	$(0, -7, 3)$	$(1, 0, -7, 3)$	$(1, 0)$
1229	$(-1, -7, 6)$	$(1, 2, -6, -1)$	$(7, 4), (0, 1), (1, 0)$
1257	$(-1, -8, 9)$	$(1, 2, -7, 1)$	$(0, 1), (1, 0)$
1300	$(0, -10, 10)$	$(1, 0, -10, 10)$	$(1, 1), (1, 0)$
1345	$(0, -7, 1)$	$(1, 0, -7, 1)$	$(-19, 7), (18, 7), (0, 1), (1, 7), (1, 0)$
1369	$(-1, -12, -11)$	$(1, 2, -11, -23)$	$(3, -1), (-2, 1), (1, 0)$
1373	$(0, -8, 5)$	$(1, 0, -8, 5)$	$(2, 3), (1, 0)$
1384	$(-1, -10, 14)$	$(1, 2, -9, 4)$	$(25, 14), (1, 2), (1, 0)$
1396	$(-1, -7, 5)$	$(1, 2, -6, -2)$	$(1, 0)$
1425	$(-1, -8, -3)$	$(1, 2, -7, -11)$	$(3, -1), (1, 0)$
1436	$(0, -11, 12)$	$(1, 0, -11, 12)$	$(5, 2), (1, 0)$
1489	$(-1, -10, -7)$	$(1, 2, -9, -17)$	$(2, -1), (3, 1), (3, -1), (-22, 7), (1, 0)$
1492	$(-1, -9, -5)$	$(1, 2, -8, -14)$	$(3, -1), (1, 0)$
1509	$(-1, -7, 4)$	$(1, 2, -6, -3)$	$(2, 1), (1, 0)$
1524	$(-1, -7, 1)$	$(1, 2, -6, -6)$	$(-1, 1), (1, 0)$
1556	$(-1, -9, 11)$	$(1, 2, -8, 2)$	$(1, 0)$
1573	$(-1, -7, 2)$	$(1, 2, -6, -5)$	$(2, 1), (-13, 18), (1, 0)$
1593	$(0, -9, 7)$	$(1, 0, -9, 7)$	$(-10, 3), (5, 2), (1, 1), (1, 0)$
1620	$(0, -12, 14)$	$(1, 0, -12, 14)$	$(1, 0)$
1708	$(-1, -8, -2)$	$(1, 2, -7, -10)$	$(1, 0)$
1765	$(-1, -11, 16)$	$(1, 2, -10, 5)$	$(2, 1), (1, 0)$
1772	$(-1, -12, 8)$	$(2, 5, -2, -3)$	$(-8, 3), (-2, 3)$
	$\omega_2 = (0, 1, 0)/1$	$\omega_3 = (0, 1, 1)/2$	
1825	$(-1, -8, 7)$	$(1, 2, -7, -1)$	$(2, 1), (0, 1), (1, 0)$
1849	$(-1, -14, -8)$	$(2, 5, -3, -8)$	no solutions
	$\omega_2 = (0, 1, 0)/1$	$\omega_3 = (0, 1, 1)/2$	
1901	$(-1, -9, -4)$	$(1, 2, -8, -13)$	$(-42, 13), (-3, 2), (1, 0)$
1929	$(-1, -10, 13)$	$(1, 2, -9, 3)$	$(2, 1), (1, 0)$
1937	$(-1, -8, -1)$	$(1, 2, -7, -9)$	$(-1, 1), (1, 0)$
1940	$(0, -8, 2)$	$(1, 0, -8, 2)$	$(-3, 1), (1, 4), (1, 0)$
1944	$(0, -9, 6)$	$(1, 0, -9, 6)$	$(1, 0)$
1957	$(-1, -9, 10)$	$(1, 2, -8, 1)$	$(-4, 1), (2, 1), (0, 1), (1, 0)$
2021	$(0, -8, 1)$	$(1, 0, -8, 1)$	$(-26, 9), (0, 1), (1, 8), (1, 0)$
2024	$(-1, -10, -6)$	$(1, 2, -9, -16)$	$(1, 0)$
2057	$(0, -11, 11)$	$(1, 0, -11, 11)$	$(1, 1), (1, 0)$
2089	$(0, -13, 4)$	$(2, 3, -5, -2)$	no solutions
	$\omega_2 = (0, 1, 0)/1$	$\omega_3 = (0, 1, 1)/2$	
2101	$(-1, -11, -8)$	$(1, 2, -10, -19)$	$(-2, 1), (1, 0)$
2177	$(-1, -8, 5)$	$(1, 2, -7, -3)$	$(2, 1), (1, 0)$
2213	$(-1, -13, -12)$	$(1, 2, -12, -25)$	$(-2, 1), (1, 0)$
2233	$(-1, -8, 1)$	$(1, 2, -7, -7)$	$(-1, 1), (1, 0)$
2241	$(0, -9, 5)$	$(1, 0, -9, 5)$	$(8, 3), (1, 0)$
2296	$(-1, -14, -14)$	$(1, 2, -13, -28)$	$(-9, 4), (1, 0)$
2300	$(-1, -8, 2)$	$(1, 2, -7, -6)$	$(-7, 2), (1, 0)$
2349	$(0, -12, 13)$	$(1, 0, -12, 13)$	$(8, 3), (1, 0)$
2505	$(-1, -10, -5)$	$(1, 2, -9, -15)$	$(1, 0)$
2557	$(-1, -9, -2)$	$(1, 2, -8, -11)$	$(1, 0)$
2589	$(-1, -14, 12)$	$(2, 5, -3, -3)$	$(1, 1)$
	$\omega_2 = (0, 1, 0)/1$	$\omega_3 = (0, 1, 1)/2$	
2597	$(-1, -9, 8)$	$(1, 2, -8, -1)$	$(2, 1), (-4, 1), (0, 1), (1, 0)$
2636	$(0, -14, 4)$	$(2, 0, -7, 1)$	$(-2, 1), (0, 1)$
	$\omega_2 = (0, 1, 0)/1$	$\omega_3 = (0, 0, 1)/2$	
2673	$(0, -9, 3)$	$(1, 0, -9, 3)$	$(1, 3), (1, 0)$
2677	$(0, -10, 7)$	$(1, 0, -10, 7)$	$(1, 0)$
2700	$(0, -15, 20)$	$(1, 0, -15, 20)$	$(1, 0)$
2708	$(-1, -11, -7)$	$(1, 2, -10, -18)$	$(1, 0)$
2713	$(0, -13, 15)$	$(1, 0, -13, 15)$	$(-49, 12), (4, 3), (1, 0)$
2777	$(-1, -14, 23)$	$(1, 2, -13, 9)$	$(-5, 1), (2, 1), (17, 8), (1, 1), (1, 0)$
2804	$(-1, -9, -1)$	$(1, 2, -8, -10)$	$(-1, 1), (1, 0)$
2808	$(0, -9, 2)$	$(1, 0, -9, 2)$	$(1, 0)$
2836	$(-1, -9, 7)$	$(1, 2, -8, -2)$	$(1, 0)$
2857	$(-1, -10, 11)$	$(1, 2, -9, 1)$	$(-21, 5), (2, 1), (1, 0), (0, 1)$
2917	$(-1, -13, 20)$	$(1, 2, -12, 7)$	$(2, 1), (1, 0)$
2981	$(-1, -11, 14)$	$(1, 2, -10, 3)$	$(2, 1), (1, 0)$
2993	$(-1, -12, 17)$	$(1, 2, -11, 5)$	$(2, 1), (1, 2), (1, 0)$
3021	$(-1, -9, 6)$	$(1, 2, -8, -3)$	$(1, 0)$
3028	$(0, -10, 6)$	$(1, 0, -10, 6)$	$(1, 0)$
3137	$(0, -11, 9)$	$(1, 0, -11, 9)$	$(-11, 3), (1, 1)$

11.1.2 Complex cubic fields

D_K	$f(x)$	$I(x, y)$	solutions
-23	$(0, -1, 1)$	$(1, 0, -1, 1)$	$(-1, 1), (0, 1), (1, 0), (1, 1), (4, -3)$
-31	$(0, 1, 1)$	$(1, 0, 1, 1)$	$(-2, 3), (0, 1), (1, -1), (1, 0)$
-44	$(1, -1, 1)$	$(1, -2, 0, 2)$	$(-47, 56), (1, -1), (1, 0), (1, 1)$
-59	$(0, 2, 1)$	$(1, 0, 2, 1)$	$(0, 1), (1, -2), (1, 0)$
-76	$(0, -2, 2)$	$(1, 0, -2, 2)$	$(-23, 13), (1, 0), (1, 1)$
-83	$(1, 1, 2)$	$(1, -2, 2, 1)$	$(0, 1), (1, 0)$
-87	$(1, -2, -3)$	$(1, -2, -1, -1)$	$(0, -1), (1, 0)$
-104	$(0, -1, 2)$	$(1, 0, -1, 2)$	$(-3, 2), (1, 0)$
-107	$(1, 3, 2)$	$(1, -2, 4, -1)$	$(0, -1), (1, 0)(2, 7)$
-108	$(0, 0, 2)$	$(1, 0, 0, 2)$	$(-1, 1), (1, 0)$
-116	$(1, 0, 2)$	$(1, -2, 1, 2)$	$(1, 0)$
-135	$(0, -3, 3)$	$(1, 0, -3, 3)$	$(-2, 1), (1, 0), 1, 1)$
-139	$(1, 1, -2)$	$(1, -2, 2, -3)$	$(1, 0), (2, 1)$
-140	$(0, 2, 2)$	$(1, 0, 2, 2)$	$(1, -1), (1, 0)$
-152	$(1, -2, 2)$	$(1, -2, -1, 4)$	$(1, 0)$
-172	$(1, -1, -3)$	$(1, -2, 0, -2)$	$(1, 0)$
-175	$(1, 2, 3)$	$(1, -2, 3, 1)$	$(0, 1), (1, 0)$
-199	$(1, -4, 3)$	$(1, -2, -3, 7)$	$(1, 0), (2, 1)$
-200	$(1, 2, -2)$	$(1, -2, 3, -4)$	$(1, 0)$
-204	$(1, 1, 3)$	$(1, -2, 2, 2)$	$(1, 0)$
-211	$(0, -2, 3)$	$(1, 0, -2, 3)$	$(1, 0), (2, -1)$
-212	$(1, 4, 2)$	$(1, -2, 5, -2)$	$(1, 0), (1, 2)$
-216	$(0, 3, 2)$	$(1, 0, 3, 2)$	$(1, 0)$
-231	$(1, 0, -3)$	$(1, -2, 1, -3)$	$(-2, -1), (1, 0)$
-239	$(0, -1, 3)$	$(1, 0, -1, 3)$	$(-5, 3), (1, 0)$
-243	$(0, 0, 3)$	$(1, 0, 0, 3)$	$(1, 0)$
-244	$(1, -4, -6)$	$(1, -2, -3, -2)$	$(1, 0)$
-247	$(0, 1, 3)$	$(1, 0, 1, 3)$	$(-1, 1), (1, 0)$
-255	$(1, 0, 3)$	$(1, -2, 1, 3)$	$(1, -1), (1, 0)$
-268	$(1, -3, -5)$	$((1, -2, -2, -2)$	$(1, 0), (3, 1)$
-283	$(0, 4, 1)$	$(1, 0, 4, 1)$	$(0, 1), (1, -4), (1, 0)$
-300	$(1, -3, 3)$	$(1, -2, -2, 6)$	$(1, 0)$

11.2 Quartic fields

The input data of quartic fields used in this Section were taken from the tables of J. Buchmann and D. Ford [BF89], D. Ford [Fo91] and J. Buchmann, D. Ford and M. Pohst [BFP93].

We present two tables that describe the behavior of the minimal indices of quartic fields. Computations were performed in various Galois groups (C_4 cyclic, D_4 dihedral, V_4 Klein four group, A_4 alternating group, S_4 symmetric group) and signatures ("r" totally real, "c" totally complex) up to discriminant $|D_K| < 10^6$.

Some of the data were computed by using the method of I. Gaál, A. Pethő and M. Pohst [GPP93] for finding "small solutions" $\max(|x_2|, |x_3|, |x_4|) < 10^{10}$ of index form equations

$$I(x_2, x_3, x_4) = \pm m$$

in quartic fields. The values we obtained for m are very likely the exact minimal indices. For the Galois group V_4 the results were computed by the algorithm of Section 6.5 and for totally complex quartic fields by the method of cf. Section 6.4.

From a statistical point of view the most interesting is to consider the data of totally real quartic fields with Galois group D_4 and especially of totally complex quartic fields with Galois group S_4. (We remark that our programs did not terminate in reasonable computation time for about 1.5% of the later examples because of extreme input data. These fields are not included in Tables 11.1, 11.2.)

11.2.1   The distribution of the minimal indices

In Table 11.1 the first row contains the Galois groups and the signature ("r" or "c"), the second row the number of fields K with the given Galois group, signature and discriminant $< 10^6$. The rows 3–8 contain intervals for the minimal index $m = m(K)$, and indicate what percentage of the fields have minimal index in the given interval. Finally, row 9 contains the maximum of the minimal indices obtained for that set of fields, and row 10 the discriminant of the field with the maximal m.

Table 11.1: The distribution of the minimal indices

	$A_4(r)$	$C_4(r)$	$V_4(r)$	$D_4(r)$	$A_4(c)$	$S_4(c)$
#K (100%)	31	59	196	4486	90	44122
$m = 1$	61.3%	14.6%	4.0%	13.8%	34.4%	26.7%
$2 \le m \le 5$	29.0%	18.2%	19.9%	24.7%	32.2%	41.6%
$6 \le m \le 15$	9.7%	16.3%	23.0%	25.7%	28.9%	27.2%
$16 \le m \le 50$	0%	31.0%	35.7%	29.4%	4.4%	4.4%
$51 \le m \le 100$	0%	16.3%	11.7%	5.3%	0%	0.002%
$m > 100$	0%	3.6%	5.6%	1.0%	0%	0%
max m	10	236	192	196	27	54
$D(\max m)$	505521	882000	970225	973225	962361	851724

We see that most fields have a small minimal index ($m \le 5$) and only very few fields have $m > 50$. Especially for $A_4(r)$, $A_4(c)$ and $S_4(c)$ there are relatively many fields with $m = 1$, that is with a power integral basis. The behavior of m is more extreme in cases $C_4(r)$, $V_4(r)$, and $D_4(r)$ than in the other cases. We found the largest m for a $C_4(r)$ field.

11.2.2   The average behavior of the minimal indices

In Table 11.2 the first column gives intervals for the discriminants of the fields. The first row contains Galois groups and signatures. The other entries of the table show the average of the minimal indices obtained for the fields with properties indicated by the row and column headings. In the last row we list the average of the minimal indices for all fields with given Galois group, signature and discriminant $< 10^6$.

Especially, the columns $S_4(c)$ and $D_4(r)$ show that the average of the minimal indices increases almost monotonically if we increase the discriminant. The total averages for $C_4(r)$ and $V_4(r)$ are much larger than the other ones, although these fields are considered as the ones with the nicest properties.

Table 11.2: The average behavior of the minimal indices

D			$A_4(r)$	$C_4(r)$	$V_4(r)$	$D_4(r)$	$A_4(c)$	$S_4(c)$
$0 <$	D	< 100000	1.7	4.9	7.7	4.1	2.7	2.2
$100000 \leq$	D	< 200000	2.0	21.5	20.9	8.0	3.6	3.3
$200000 \leq$	D	< 300000	1.2	13.4	29.1	11.0	3.8	3.8
$300000 \leq$	D	< 400000	1.0	41.1	21.2	13.3	7.0	4.4
$400000 \leq$	D	< 500000	3.2	46.5	35.9	15.9	3.7	4.8
$500000 \leq$	D	< 600000	3.0	50.3	37.4	17.9	8.8	5.2
$600000 \leq$	D	< 700000	1.0	59.0	39.3	20.0	6.0	5.5
$700000 \leq$	D	< 800000	9.0	19.0	45.2	22.5	4.7	5.7
$800000 \leq$	D	< 900000	2.3	73.8	39.1	25.0	5.8	5.9
$900000 \leq$	D	< 1000000	4.5	36.7	61.3	25.2	12.4	6.0
Total		average	2.5	28.5	28.7	16.5	4.8	4.8

11.2.3 Totally real cyclic quartic fields

The fields below are totally real cyclic quartic fields of the form $K = \mathbb{Q}(\sqrt{\mu})$ with smallest discriminants D_K, integral bases $\{1, \omega, \psi, \omega\psi\}$. We list D_K, μ, ω, ψ and the solutions (x_2, x_3, x_4) of the index form equation corresponding to this integral basis. The data were computed by the method of I. Gaál, A. Pethő and M. Pohst [GPP91b].

$D_K = 1125$ $\mu = 15 + 6\sqrt{5}, \ \omega = (1 + \sqrt{5})/2, \ \psi = (1 + \sqrt{\mu})/2$
$(-9, -13, 8), (-6, 1, 2), (-4, -1, -2), (-2, -3, 2),$
$(-1, -2, 1), (-1, -1, 0), (-1, -1, 1), (-1, 1, 0),$
$(-1, 13, -8), (0, -3, 2), (0, -2, 1), (0, -1, 1)$

$D_K = 2000$ $\mu = 5 + 2\sqrt{5}, \ \omega = (1 + \sqrt{5})/2, \ \psi = \sqrt{\mu}$
$(-2, -5, 3), (-2, 0, -1), (-2, 0, 1), (-2, 5, -3),$
$(-1, -3, 2), (-1, -1, 0), (-1, 1, 0), (-1, 3, -2),$
$(0, -2, 1), (0, -1, 1)$

$D_K = 2048$ $\mu = 2 + \sqrt{2}, \ \omega = \sqrt{2}, \ \psi = \sqrt{\mu}$
$(-1, -1, 0), (-1, -1, 1), (-1, 1, -1), (-1, 1, 0)$
$(0, -1, 0), (0, -1, 1)$

$D_K = 6125$ $\mu = 35 + 14\sqrt{5}, \ \omega = (1 + \sqrt{5})/2, \ \psi = (1 + \sqrt{\mu})/2$
$(-1, -2, 1), (-1, -1, 1), (0, -2, 1), (0, -1, 1)$

$D_K = 8000$ $\mu = 10 + 4\sqrt{5}, \ \omega = (1 + \sqrt{5})/2, \ \psi = \sqrt{\mu}$
no solutions

$D_K = 15125$ $\mu = 55 + 22\sqrt{5}, \ \omega = (1 + \sqrt{5})/2, \ \psi = (1 + \sqrt{\mu})/2$
$(-1, -2, 1), (-1, -1, 1), (0, -2, 1), (0, -1, 1)$

$D_K = 18432$ $\mu = 6 + 3\sqrt{2}, \ \omega = \sqrt{2}, \ \psi = \sqrt{\mu}$
no solutions

$D_K = 45125$ $\mu = 95 + 38\sqrt{5}, \ \omega = (1 + \sqrt{5})/2, \ \psi = (1 + \sqrt{\mu})/2$
no solutions

$D_K = 51200$ $\mu = 10 + 5\sqrt{2}, \ \omega = \sqrt{2}, \ \psi = \sqrt{\mu}$
no solutions

$D_K = 66125$ $\mu = 115 + 46\sqrt{5}, \ \omega = (1 + \sqrt{5})/2, \ \psi = (1 + \sqrt{\mu})/2$
no solutions

$D_K = 72000$ $\mu = 15 + 3\sqrt{5}, \ \omega = (1 + \sqrt{5})/2, \ \psi = \sqrt{\mu}$
no solutions

$D_K = 120125$ $\mu = 82 + 22\sqrt{5}, \ \omega = (1 + \sqrt{5})/2, \ \psi = \sqrt{\mu}$
no solutions

$D_K = 190125$ $\mu = 195 + 78\sqrt{5}, \ \omega = (1 + \sqrt{5})/2, \ \psi = (1 + \sqrt{\mu})/2$
no solutions

11.2.4 Monogenic mixed dihedral extensions of real quadratic fields

To illustrate our algorithm described in Section 6.3 we performed computations for $M = \mathbb{Q}(\sqrt{2})$, $\mathbb{Q}(\sqrt{3})$, $\mathbb{Q}(\sqrt{5})$. Our purpose is to determine all mixed dihedral extensions of M that are monogenic.

The following table contains the results of our computation. For each of these quadratic fields M we list

$$D_K, \ (a_1, a_2, a_3, a_4), \ d, \ (x, y, z), \ldots$$

where D_K is the discriminant of the mixed dihedral quartic field K containing M as a subfield and having power integral bases; $f(x) = x^4 + a_1x^3 + a_2x^2 + a_3x + a_4$ is the minimal polynomial of the generating element ξ of K. The integral elements of K are represented in the form

$$\alpha = \frac{a + x\xi + y\xi^2 + z\xi^3}{d} \tag{11.1}$$

with integer a, x, y, z and with a common denominator $0 < d \in \mathbb{Z}$, also included in the table. Finally we display the coordinates (x, y, z) of the α in (11.1) generating power integral bases.

At some fields *no solutions* means that the field K is a mixed dihedral quartic field containing M as a subfield, but having no power integral bases.

$M = \mathbb{Q}(\sqrt{2})$

D_K	$f(x)$	d	solutions
-448	$(-2, 1, 2, -1)$	1	$(0, 1, -1), (1, -6, 4), (4, -3, 1), (1, 0, 0), (1, -2, 1)$
-1024	$(0, -2, 0, -1)$	1	$(1, 0, 0), (2, 0, -1)$
-1472	$(-2, -3, -2, -1)$	1	$(1, 0, 0), (0, 3, -1), (3, 2, -1)$
-1792	$(0, -2, -4, -2)$	1	$(1, 0, 0), (1, 1, -1)$
-1984	$(-2, 1, 0, -2)$	1	$(1, 0, 0)$
-2048	$(0, 0, 0, -2)$	1	$(1, 0, 0), (1, 1, 1), (1, -1, 1)$

$M = \mathbb{Q}(\sqrt{3})$

D_K	$f(x)$	d	solutions
-1728	$(-2, 0, -2, 1)$	1	$(1, 0, 0), (0, 2, -1)$
-3312	$(-2, -1, 2, -2)$	1	$(1, 0, 0)$
-3312	$(0, 1, -6, 1)$	3	no solutions
-4608	$(0, 2, 0, -2)$	1	$(1, 0, 0), (3, 1, 1), (3, -1, 1)$
-4608	$(0, -2, 0, -2)$	1	$(1, 0, 0)$
-5616	$(0, -3, -6, -3)$	1	$(2, 1, -1), (1, 0, 0)$
-5616	$(-2, 3, -2, -2)$	1	$(1, 0, 0), (17, -10, 4), (9, -2, 4)$
-6336	$(-2, -4, -4, -2)$	1	$(1, 0, 0), (1, 3, -1)$
-6336	$(-2, -8, -6, -3)$	3	no solutions
-6768	$(-2, 1, 0, -3)$	1	$(1, 0, 0), (1, 0, 1), (4, -3, 1)$
-6768	$(0, -1, -6, -2)$	2	$(2, 1, -2), (2, -1, 0), (2, 1, 0), (1, 1, -1)$
-6912	$(0, 0, 0, -3)$	1	$(1, 0, 0)$

$M = \mathbb{Q}(\sqrt{5})$

D_K	$f(x)$	d	solutions
-275	$(-1, 0, 2, -1)$	1	$(0, 0, 1), (1, 0, 0), (2, -2, 1), (1, 2, -4), (0, 1, -1)$
-400	$(0, -1, 0, -1)$	1	$(1, 0, 0), (0, 1, 1), (1, 0, -1), (0, 1, -1)$
-475	$(-1, -2, -2, -1)$	1	$(1, 0, 0), (0, 2, -1), (2, 1, -1)$

11.2.5   Totally real bicyclic biquadratic number fields

Using the algorithm described in Section 6.5.2 we determined the minimal index and all elements with minimal index in the 196 totally real bicyclic biquadratic number fields $K = \mathbb{Q}(\sqrt{m}, \sqrt{n})$ with discriminant $< 10^6$. According to the notation of Section 6.5 in our table, $l = (m, n)$, and m_1, n_1 is defined by $m = lm_1, n = ln_1$. The discriminant of K is D_K, the field index (computed by the algorithm of Section 6.5.4) is $m(K)$ and the minimal index is $\mu(K)$. We use the integral basis given in Section 6.5.1 depending on the remainders of m, n, m_1, n_1 modulo 4. The solutions (x_2, x_3, x_4) refer to the index form equation corresponding to this integral basis with right hand side $\mu(K)$.

D_K	n_1	m_1	l	$m(K)$	$\mu(K)$	(x_2, x_3, x_4)
1600	2	5	1	1	3	$(-7, -8, 5), (-7, 3, 5), (-1, -1, 1), (-1, 0, 1),$
						$(-1, 1, 0), (1, -1, 1), (1, 0, 1), (1, 1, 0), (7, -8, 5),$
						$(7, 3, 5)$
2304	3	2	1	1	1	$(-4, -2, 3), (-4, 2, 3), (-1, 0, 1), (0, 0, 1),$
						$(1, -2, 3), (1, 2, 3)$
3600	1	5	3	1	4	$(-1, 1, -1), (0, -1, -1), (0, 1, -1), (1, -1, -1)$
4225	5	13	1	2	6	$(-2, -3, 1), (-2, 2, 1), (-1, -1, 1), (-1, 0, 1),$
						$(0, -1, 1), (0, 0, 1), (1, -3, 1), (1, 2, 1)$
7056	3	7	1	1	1	$(-1, 1, 0), (0, 1, 0)$
7225	5	17	1	2	12	$(-5, -8, 3), (-5, 5, 3), (-3, -5, 2), (-3, 3, 2),$
						$(-1, -1, 1), (-1, 0, 1), (-1, 1, 0), (-1, 2, 0),$
						$(0, -1, 1), (0, 0, 1), (1, -5, 2), (1, 1, 0), (1, 2, 0),$
						$(1, 3, 2), (2, -8, 3), (2, 5, 3)$
10816	2	13	1	1	5	$(-1, -2, 1), (-1, 1, 0), (-1, 1, 1), (1, -2, 1),$
						$(1, 1, 0), (1, 1, 1)$
11025	5	21	1	2	4	$(-1, 2, 0), (1, 2, 0)$
12544	7	2	1	1	3	$(-1, 0, 1), (0, 0, 1)$
14400	6	5	1	1	5	$(-1, -1, 1), (-1, 0, 1), (1, -1, 1), (1, 0, 1)$
17424	3	11	1	1	2	$(-1, 1, 0), (0, 1, 0)$
18496	2	17	1	1	2	$(-1, -2, 1), (-1, 1, 1), (1, -2, 1), (1, 1, 1)$
19600	1	5	7	1	12	$(-8, 11, 13), (-3, 11, 13), (-1, 1, -1),$
						$(0, -1, -1), (0, 1, -1), (1, -1, -1), (3, -11, 13),$
						$(8, -11, 13)$
21025	5	29	1	2	20	$(-2, -4, 1), (-2, 3, 1), (-1, -2, 1), (-1, 1, 1),$
						$(0, -2, 1), (0, 1, 1), (1, -4, 1), (1, 3, 1)$
24336	1	13	3	1	1	$(-1, 2, 1), (1, -2, 1)$
27225	5	33	1	2	18	$(-1, -2, 1), (-1, 1, 1), (0, -2, 1), (0, 1, 1)$
28224	2	21	1	1	9	$(-1, -2, 1), (-1, 1, 1), (1, -2, 1), (1, 1, 1)$
28224	2	7	3	1	4	$(0, -1, 1), (0, 0, 1)$
30976	11	2	1	1	5	$(-1, 0, 1), (0, 0, 1)$
34225	5	37	1	2	14	$(-5, -11, 3), (-5, 8, 3), (-1, -2, 1), (-1, 1, 1),$
						$(0, -2, 1), (0, 1, 1), (2, -11, 3), (2, 8, 2)$
41616	1	17	3	1	5	$(-1, 2, 1), (1, -2, 1)$
42025	5	41	1	2	8	$(-1, -2, 1), (-1, 1, 1), 0, -2, 1), (0, 1, 1)$
48400	1	5	11	1	9	$(-1, 2, 3), (1, -2, 3)$
48841	13	17	1	2	4	$(-1, 1, 0), (1, 1, 0)$
51984	3	19	1	1	4	$(-1, 1, 0), (0, 1, 0)$
53361	7	11	3	2	2	$(0, -1, 1), (0, 0, 1), (1, -1, 1), (1, 0, 1)$
53824	2	29	1	1	12	$(-1, 2, 0), (1, 2, 0)$
57600	15	2	1	1	7	$(-7, -2, 3), (-7, 2, 3), (-1, 0, 1), (0, 0, 1),$
						$(4, -2, 3), (4, 2, 3)$
57600	3	10	1	1	3	$(-1, -1, 1), (-1, 1, 1), (0, -1, 1), (0, 1, 1)$
57600	5	2	3	1	1	$(-1, 0, 1), (0, 0, 1)$
69696	2	33	1	1	4	$(-1, 2, 0), (1, 2, 0)$
69696	2	11	3	1	8	$(0, -1, 1), (0, 0, 1)$
70225	5	53	1	2	22	$(-1, -2, 1), (-1, 1, 1), (0, -2, 1), (0, 1, 1)$
74529	13	21	1	2	8	$(-1, 1, 0), (1, 1, 0)$
76176	3	23	1	1	5	$(-1, 1, 0), (0, 1, 0)$
78400	14	5	1	1	11	$(-5, -1, -1), (-5, 2, -1), (5, -1, -1),$
						$(5, 2, -1)$
81225	5	57	1	2	36	$(-1, -2, 1), (-1, 1, 1), (0, -2, 1), (0, 1, 1)$

D_K	n_1	m_1	l	$m(K)$	$\mu(K)$	(x_2, x_3, x_4)
87616	2	37	1	1	20	$(-1, 2, 0), (1, 2, 0)$
92416	19	2	1	1	9	$(-1, 0, 1), (0, 0, 1)$
93025	5	61	1	2	12	$(-3, -8, 2), (-3, 6, 2), (1, -8, 2), (1, 6, 2)$
94864	7	11	1	1	1	$(-9, 10, 3), (-1, 1, 0), (-1, 10, 3), (0, 1, 0),$ $(1, -10, 3), (9, -10, 3)$
97344	6	13	1	1	4	$(-2, -2, 1), (-2, 1, 1), (2, -1, 1), (2, 1, 1)$
107584	2	41	1	1	33	$(-1, 1, 0), (1, 1, 0)$
112896	3	14	1	1	5	$(-1, -1, 1), (-1, 1, 1), (0, -1, 1), (, 0, 1, 1)$
112896	7	6	1	1	8	$(-1, 1, 0), (1, 1, 0)$
119025	5	69	1	2	52	$(-2, -6, 1), (-2, 5, 1), (1, -6, 1), (1, 5, 1)$
121104	1	29	3	1	17	$(-1, 2, 1), (1, -2, 1)$
127449	17	21	1	2	4	$(-1, 1, 0), (1, 1, 0)$
132496	1	13	7	3	15	$(-7, 11, 8), (-4, 11, 8), (-1, 2, 1), (1, -2, 1),$ $(4, -11, 8), (7, -11, 8)$
133225	5	73	1	2	68	$(-1, 1, 0), (1, 1, 0)$
135424	23	2	1	1	11	$(-1, 0, 1), (0, 0, 1)$
138384	3	31	1	1	7	$(-1, 1, 0), (0, 1, 0)$
142129	13	29	1	2	16	$(-2, -3, 1), (-2, 2, 1), (-1, 1, 0), (1, -3, 1),$ $(1, 1, 0), (1, 2, 1)$
144400	1	5	19	1	15	$(-1, 1, 2), (0, -1, 2), (0, 1, 2), (1, -1, 2)$
148225	5	77	1	2	48	$(-1, 4, 0), (1, 4, 0)$
159201	7	19	3	2	4	$(-1, -7, 4), (-1, 3, 4), (5, -7, 4), (5, 3, 4)$
166464	6	17	1	1	7	$(-1, 1, 0), (1, 1, 0)$
176400	3	35	1	1	8	$(-4, 6, 1), (-2, 6, 1), (-1, 1, 0), (0, 1, 0), (2, -6, 1),$ $(4, -6, 1)$
176400	7	15	1	1	2	$(-1, 1, 0), (0, 1, 0)$
176400	3	7	5	1	4	$(-1, 1, -1), (0, -1, -1), (0, 1, -1), (1, -1, -1)$
179776	2	53	1	1	21	$(-1, -3, 1), (-1, 2, 1), (1, -3, 1), (1, 2, 1)$
184041	13	33	1	2	14	$(-2, -3, 1), (-2, 2, 1), (1, -3, 1), (1, 2, 1)$
193600	22	5	1	1	36	$(-2, -1, 1), (-2, 0, 1), (2, -1, 1), (2, 0, 1)$
197136	1	37	3	1	3	$(-4, 7, 2), (-3, 7, 2), (3, -7, 2), (4, -7, 2)$
198025	5	89	1	2	84	$(-1, 1, 0), (1, 1, 0)$
207936	2	57	1	1	7	$(-3, -9, 2), (-3, 7, 2), (3, -9, 2), (3, 7, 2)$
207936	2	19	3	1	2	$(-1, -1, -1), (-1, 2, -1), (1, -1, 1), (1, 2, -1)$
211600	1	5	23	1	36	$(-5, 7, 15), (-2, 7, 15), (2, -7, 15), (5, -7, 15)$
216225	5	93	1	2	38	$(-2, -7, 1), (-2, 6, 1), (1, -7, 1), (1, 6, 1)$
226576	1	17	7	1	11	$(-1, 2, 1), (1, -2, 1),$
231361	13	37	1	6	6	$(-2, -3, 1), (-2, 2, 1), (1, -3, 1), (1, 2, 1)$
233289	7	23	3	2	10	$(0, -2, 1), (0, 1, 1), (1, -2, 1), (1, 1, 1)$
235225	5	97	1	2	92	$(-1, 1, 0), (1, 1, 0)$
238144	2	61	1	1	53	$(-7, -22, 5), (-7, 17, 5), (-1, 1, 0), (1, 1, 0),$ $(7, -22, 5), (7, 17, 5)$
242064	1	41	3	1	28	$(-2, 4, 1), (2, -4, 1)$
243049	17	29	1	2	12	$(-1, 1, 0), (1, 1, 0)$
246016	31	2	1	1	15	$(-1, 0, 1), (0, 0, 1)$
255025	5	101	1	2	20	$(-3, -10, 2), (-3, 8, 2), (1, -10, 2), (1, 8, 2)$
266256	3	43	1	1	10	$(-1, 1, 0), (0, 1, 0)$
270400	2	65	1	1	57	$(-1, 1, 0), (1, 1, 0)$
270400	26	5	1	1	99	$(-5, -1, 1), (-5, 0, 1), (-1, 1, 0), (1, 1, 0),$ $(5, -1, 1), (5, 0, 1)$
270400	10	13	1	3	27	$(-3, -2, 1), (-3, 1, 1), (-1, 1, 0), (1, 1, 0),$ $(3, -2, 1), (3, 1, 1)$
270400	2	13	5	1	8	$(-2, -4, 3), (-2, 1, 3), (0, -1, 1), (0, 0, 1),$ $(2, -4, 3), (2, 1, 3)$
278784	3	22	1	1	19	$(-4, -7, 3), (-4, 7, 3), (1, -7, 3), (1, 7, 3)$
278784	11	6	1	1	5	$(-2, -1, 1), (-2, 1, 1), (1, -1, 1), (1, 1, 1)$
283024	7	19	1	3	3	$(-1, 1, 0), (0, 1, 0)$
284089	13	41	1	2	28	$(-23, -37, 10), (-23, 27, 10), (-1, 1, 0), (1, 1, 0),$ $(13, -37, 10), (13, 27, 10)$
297025	5	109	1	2	84	$(-1, -3, 1), (-1, 2, 1), (0, -3, 1), (0, 2, 1)$
304704	2	69	1	1	27	$(-1, 3, 0), (1, 3, 0)$

D_K	n_1	m_1	l	$m(K)$	$\mu(K)$	(x_2, x_3, x_4)
304704	2	23	3	1	5	$(-1, -1, -1), (-1, 2, -1), (1, -1, -1), (1, 2, -1)$
313600	35	2	1	1	17	$(-1, 0, 1), (0, 0, 1)$
313600	7	10	1	3	24	$(-1, 1, 0), (1, 1, 0)$
313600	7	2	5	1	1	$(-1, 0, 1), (0, 0, 1)$
314721	17	33	1	4	16	$(-2, -3, -1), (-2, 4, -1), (-1, 1, 0), (1, 1, 0),$
						$(3, -3, -1), (3, 4, -1)$
318096	3	47	1	1	11	$(-1, 1, 0), (0, 1, 0)$
319225	5	113	1	2	66	$(-1, -3, 1), (-1, 2, 1), (0, -3, 1), (0, 2, 1)$
327184	1	13	11	1	31	$(-7, 11, 10), (-4, 11, 10), (-1, 2, 1), (1, -2, 1),$
						$(4, -11, 10), (7, -11, 10)$
341056	2	73	1	1	9	$(-1, 3, 0), (1, 3, 0)$
370881	21	29	1	2	8	$(-14, -16, 5), (-14, 11, 5), (-1, 1, 0), (1, 1, 0),$
						$(9, -16, 5), (9, 11, 5)$
379456	2	77	1	1	45	$(-1, 3, 0), (1, 3, 0)$
379456	2	11	7	1	4	$(0, -1, 1), (0, 0, 1)$
384400	1	5	31	1	25	$(-1, 2, 5), (1, -2, 5)$
389376	39	2	1	1	19	$(-1, 0, 1), (0, 0, 1)$
389376	3	26	1	1	55	$(-15, -28, 11), (-15, 28, 11), (-1, -2, 1), (-1, 2, 1),$
						$(0, -2, 1), (0, 2, 1), (4, -28, 11), (4, 28, 11)$
389376	13	2	3	1	5	$(-1, 0, 1), (0, 0, 1)$
393129	11	19	3	4	16	$(0, -1, 1), (0, 0, 1), (1, -1, 1), (1, 0, 1)$
395641	17	37	1	2	20	$(-5, -7, 2), (-5, 5, 2), (-1, 1, 0), (1, 1, 0),$
						$(3, -7, 2), (3, 5, 2)$
404496	1	53	3	1	20	$(-2, 4, 1), (2, -4, 1)$
414736	7	23	1	1	4	$(-1, 1, 0), (0, 1, 0)$
416025	5	129	1	2	26	$(-1, -3, 1), (-1, 2, 1), (0, -3, 1), (0, 2, 1)$
423801	7	31	3	2	8	$(0, -2, 1), (0, 1, 1), (1, -2, 1), (1, 1, 1)$
435600	3	55	1	1	13	$(-1, 1, 0), (, 0, 1, 0)$
435600	11	15	1	1	1	$(-1, 1, 0), (0, 1, 0)$
435600	3	11	5	1	24	$(-1, 1, -1), (0, -1, -1), (0, 1, -1), (1, -1, -1)$
442225	5	133	1	2	54	$(-1, -3, 1), (-1, 2, 1), (0, -3, 1), (0, 2, 1)$
462400	2	85	1	1	77	$(-1, -5, 1), (-1, 1, 0), (-1, 4, 1), (1, -5, 1),$
						$(1, 1, 0), (1, 4, 1)$
462400	34	5	1	1	131	$(-1, 1, 0), (1, 1, 0)$
462400	10	17	1	1	23	$(-1, 1, 0), (1, 1, 0)$
462400	2	17	5	3	12	$(-12, -27, 19), (-12, 8, 19), (0, -1, 1), (0, 0, 1),$
						$(12, -27, 19), (12, 8, 19)$
469225	5	137	1	2	84	$(-1, -3, 1), (-1, 2, 1), (0, -3, 1), (0, 2, 1)$
473344	43	2	1	1	21	$(-1, 0, 1), (0, 0, 1)$
474721	13	53	1	2	4	$(-1, 2, 0), (1, 2, 0)$
484416	6	29	1	1	5	$(-1, 1, 0), (1, 1, 0)$
485809	17	41	1	4	24	$(-1, 1, 0), (1, 1, 0)$
497025	5	141	1	2	116	$(-1, -3, 1), (-1, 2, 1), (0, -3, 1), (0, 2, 1)$
501264	3	59	1	1	14	$(-1, 1, 0), (0, 1, 0)$
506944	2	89	1	1	11	$(-3, -11, 2), (-3, 9, 2), (3, -11, 2), (3, 9, 2)$
529984	14	13	1	1	16	$(-2, 1, 0), (2, 1, 0)$
535824	1	61	3	1	5	$(-5, 9, 2), (-4, 9, 2), (4, -9, 2), (5, -9, 2)$
549081	13	57	1	2	20	$(-7, -13, 3), (-7, 10, 3), (-1, 2, 0), (1, 2, 0),$
						$(4, -13, 3), (4, 10, 3)$
553536	2	93	1	1	55	$(-1, -4, 1), (-1, 3, 1), (1, -4, 1), (1, 3, 1)$
553536	2	31	3	1	13	$(-1, -1, -1), (-1, 2, -1), (1, -1, -1), (1, 2, -1)$
555025	5	149	1	2	144	$(-8, -33, 5), (-8, 28, 5), (-1, 1, 0), (1, 1, 0),$
						$(3, -33, 5), (3, 28, 5)$
559504	1	17	11	1	27	$(-1, 2, 1), (1, -2, 1)$
565504	47	2	1	1	23	$(-1, 0, 1), (0, 0, 1)$
576081	11	23	3	2	28	$(-1, -4, 2), (-1, 2, 2), (3, -4, 2), (3, 2, 2)$
577600	38	5	1	1	93	$(-11, -3, 2), (-11, 1, 2), (11, -3, 2), (11, 1, 2)$
602176	2	97	1	1	12	$(-1, -4, 1), (-1, 3, 1), (1, -4, 1), (1, 3, 1)$
603729	21	37	1	2	16	$(-67, -85, 24), (-67, 61, 24), (-1, 1, 0), (1, 1, 0),$
						$(43, -85, 24), (43, 61, 24)$
608400	1	65	3	1	53	$(-1, 2, 1), (1, -2, 1)$
608400	5	13	3	1	18	$(-1, 1, 0), (0, 1, 0)$
608400	1	13	15	1	32	$(-1, 2, 2), (1, -2, 2)$
608400	1	5	39	1	151	$(-1, 2, 1), (1, -2, 1)$

D_K	n_1	m_1	l	$m(K)$	$\mu(K)$	(x_2, x_3, x_4)
616225	5	157	1	2	152	$(-1, 1, 0), (1, 1, 0)$
628849	13	61	1	6	36	$(-23, -44, 10), (-23, 34, 10), (-1, 2, 0), (1, 2, 0),$
						$(13, -44, 10), (13, 34, 10)$
646416	3	67	1	1	16	$(-1, 1, 0), (0, 1, 0)$
648025	5	161	1	2	32	$(-2, -9, 1), (-2, 8, 1), (1, -9, 1), (1, 8, 1)$
652864	2	101	1	1	39	$(-1, -4, 1), (-1, 3, 1), (1, -4, 1), (1, 3, 1)$
659344	1	29	7	1	1	$(-1, 2, 1), (1, -2, 1)$
665856	51	2	1	1	25	$(-1, 0, 1), (0, 0, 1)$
665856	3	34	1	1	37	$(-1, -3, 1), (-1, 3, 1), (0, -3, 1), (0, 3, 1)$
665856	17	2	3	1	7	$(-1, 1, 0), (0, 0, 1)$
698896	11	19	1	1	2	$(-1, 1, 0), (0, 1, 0)$
705600	2	105	1	1	97	$(-1, 1, 0), (1, 1, 0)$
705600	42	5	1	1	99	$(-3, -1, 1), (-3, 0, 1), (3, -1, 1), (3, 0, 1)$
705600	10	21	1	1	19	$(-1, 1, 0), (1, 1, 0)$
705600	2	35	1	1	32	$(0, -1, 1), (0, 0, 1)$
705600	10	7	3	1	21	$(-1, -1, 1), (-1, 0, 1), (1, -1, 1), (1, 0, 1)$
705600	2	21	5	1	16	$(0, -1, 1), (0, 0, 1)$
705600	2	15	7	1	8	$(0, -1, 1), (0, 0, 1)$
725904	3	71	1	1	17	$(-1, 1, 0), (0, 1, 0)$
739600	1	5	43	1	68	$(-1, 1, 3), (0, -1, 3), (0, 1, 3), (1, -1, 3)$
741321	21	41	1	2	2	$(-3, -4, 1), (-3, 3, 1), (2, -4, 1), (2, 3, 1)$
748225	5	173	1	2	168	$(-1, -7, 1), (-1, 1, 0), (-1, 6, 1), (0, -7, 1),$
						$(0, 6, 1), (1, 1, 0)$
753424	7	31	1	3	6	$(-1, 1, 0), (0, 1, 0)$
760384	2	109	1	1	101	$(-1, 1, 0), (1, 1, 0)$
767376	1	73	3	1	48	$(-3, 5, -1), (-2, 5, -1), (2, -5, -1), (3, -5, -1)$
774400	55	2	1	1	27	$(-12, -2, 3), (-12, 2, 3), (-1, 0, 1), (0, 0, 1),$
						$(9, -2, 3), (9, 2, 3)$
774400	11	10	1	1	8	$(-1, 1, 0), (1, 1, 0)$
774400	11	2	5	3	3	$(-1, 0, 1), (0, 0, 1)$
783225	5	177	1	2	108	$(-1, 6, 0), (1, 6, 0)$
788544	6	37	1	1	12	$(-2, -3, 1), (-2, 2, 1), (2, -3, 1), (2, 1, 1)$
804609	13	69	1	2	20	$(-2, -4, 1), (-2, 3, 1), (1, -4, 1), (1, 3, 1)$
811801	17	53	1	2	36	$(-1, 1, 0), (1, 1, 0)$
815409	7	43	3	2	20	$(0, -2, 1), (0, 1, 1), (1, -2, 1), (1, 1, 1)$
817216	2	113	1	1	105	$(-1, 1, 0), (1, 1, 0)$
819025	5	181	1	2	36	$(-1, 6, 0), (1, 6, 0)$
831744	3	38	1	1	35	$(-1, -3, 1), (-1, 3, 1), (0, -3, 1), (0, 3, 1)$
831744	19	6	1	1	55	$(-2, -1, 10, (-2, 1, 1), (1, -1, 1), (1, 1, 1)$
846400	46	5	1	1	37	$(-3, -1, 1), (-3, 0, 1), (3, -1, 1), (3, 0, 1)$
853776	1	77	3	1	52	$(-3, 5, -1), (-2, 5, -1), (2, -5, -1), (3, -5, -1)$
853776	1	33	7	1	5	$(-1, 2, 1), (1, -2, 1)$
853776	1	21	11	1	23	$(-1, 2, 1), (1, -2, 1)$
883600	1	5	47	1	76	$(-1, 1, 3), (0, -1, 3), (0, 1, 3), (1, -1, 3)$
891136	59	2	1	1	29	$(-1, 0, 1), (0, 0, 1)$
898704	3	79	1	1	19	$(-1, 1, 0), (0, 1, 0)$
900601	13	73	1	6	30	$(-2, -4, 1), (-2, 3, 1), (1, -4, 1), (1, 3, 1)$
906304	14	17	1	1	39	$(-1, 1, 0), (1, 1, 0),$
915849	29	33	1	2	4	$(-1, 1, 0), (1, 1, 0)$
931225	5	193	1	2	188	$(-1, 1, 0), (1, 1, 0)$
938961	17	57	1	4	40	$(-1, 1, 0), (1, 1, 0)$
968256	6	41	1	1	17	$(-1, 1, 0), (1, 1, 0)$
970225	5	197	1	2	192	$(-3, -15, 2), (-3, 13, 2), (-1, 1, 0), (1, -15, 2),$
						$(1, 1, 0), (1, 13, 2)$
974169	7	47	3	2	20	$(0, -2, 1), (0, 1, 1), (1, -2, 1), (1, 1, 1)$
976144	1	13	19	3	63	$(-1, 2, 1), (1, -2, 1)$
992016	3	83	1	1	20	$(-1, 1, 0), (0, 1, 0)$

11.2.6   Totally complex bicyclic biquadratic number fields

The following data were computed by the methods of Section 6.5.3. We present a list of totally complex biquadratic fields up to discriminant 10^4. Using the notation of Section 6.5 in the table, $l = (m, n)$, and m_1, n_1 is defined by $m = lm_1, n = ln_1$. The discriminant of K is D_K, the field index (computed by the algorithm of Section 6.5.4) is $m(K)$ and the minimal index is $\mu(K)$. We use the integral basis given in Section 6.5.1 depending on the remainders of m, n, m_1, n_1 modulo 4. The solutions (x_2, x_3, x_4) refer to the index form equation corresponding to this integral basis with right-hand side $\mu(K)$.

D_K	m_1	n_1	l	$m(K)$	$\mu(K)$	(x_2, x_3, x_4)
144	3	−1	1	1	1	$(1, -2, 1), (1, -1, 0), (0, 1, 0), (1, -2, -1)$
225	−3	5	1	2	2	$(0, 1, -1), (0, 0, 1), (1, 0, -1), (1, 1, -1)$
256	2	−1	1	1	1	$(0, 0, 1), (1, 0, -1)$
400	−1	−5	1	1	1	$(0, 1, 0), (1, -1, 0)$
441	−1	7	3	2	2	$(0, 1, -1), (1, 0, 1), (1, -1, 1), (0, 0, 1)$
576	−3	2	1	1	4	$(0, 1, -1), (0, 0, 1)$
576	−1	2	3	1	3	$(1, 1, -1), (1, 0, -1), (1, 0, 1), (1, -1, 1)$
784	7	−1	1	1	2	$(1, -1, 0), (0, 1, 0)$
1089	−1	11	3	2	4	$(3, -1, 2), (1, 1, -2)$
1225	−7	5	1	2	6	$(0, 1, -1), (1, 1, -1), (1, 0, -1), (0, 0, 1)$
1521	−3	13	1	2	10	$(2, 1, -1), (2, 0, -1), (1, 0, 1), (1, -1, 1)$
1600	5	−2	1	1	4	$(0, 1, -1), (0, 0, 1)$
1600	1	−2	5	1	4	$(0, 1, -1), (0, 0, 1)$
1936	11	−1	1	1	3	$(1, -1, 0), (0, 1, 0)$
2304	6	−1	1	1	5	$(0, 1, -1), (0, 1, 1), (1, -1, -1), (1, 1, -1)$
2304	−2	3	1	1	1	$(1, 0, -1), (0, 0, 1)$
2601	−3	17	1	2	20	$(0, 1, -1), (1, -1, 0), (1, 1, 0), (1, 0, -1), (1, 1, -1), (0, 0, 1)$
2704	−1	−13	1	1	3	$(0, 1, 0), (1, -1, 0)$
3025	−11	5	1	2	12	$(0, 0, 1), (0, 1, -1), (1, 1, -1), (1, 0, -1)$
3136	−7	2	1	1	8	$(0, 1, -1), (0, 0, 1)$
3136	−1	2	7	1	1	$(1, -1, 2), (1, 1, -2)$
3249	−1	19	3	2	14	$(1, 0, -1), (2, -1, 1), (1, 1, -1), (2, 0, 1)$
3600	15	−1	1	1	4	$(1, -1, 0), (0, 1, 0), (2, -4, 1), (2, -4, -1)$
3600	3	−5	1	1	2	$(1, -1, 0), (0, 1, 0)$
3600	3	−1	5	1	12	$(0, 1, -1), (1, -1, 1), (0, 1, 1), (1, -1, -1)$
4624	−1	−17	1	1	4	$(0, 1, 0), (1, -1, 0)$
4761	−1	23	3	2	8	$(2, -1, 1), (1, 0, -1), (2, 0, 1), (1, 1, -1)$
5776	19	−1	1	1	5	$(1, -1, 0), (0, 1, 0)$
5929	−1	11	7	2	4	$(2, -1, 2), (0, 1, -2)$
6400	10	−1	1	1	21	$(0, 1, -1), (1, 1, -1), (1, -1, -1), (0, 1, 1)$
6400	2	−5	1	1	3	$(1, 0, -1), (0, 0, 1)$
6400	2	−1	5	3	3	$(1, 0, -1), (0, 0, 1)$
7056	1	−7	3	1	18	$(1, -1, 0), (0, 1, 0)$
7056	1	−3	7	1	31	$(1, 0, -1), (1, 0, 1)$
7056	−1	−21	1	1	5	$(0, 1, 0), (1, -1, 0)$
7569	−3	29	1	2	26	$(3, 1, -1), (2, -1, 1), (3, 0, -1), (2, 0, 1)$
7744	−11	2	1	1	12	$(0, 0, 1), (0, 1, -1)$
7744	−1	2	11	3	3	$(1, -1, 2), (1, 1, -2)$
8281	−7	13	1	2	20	$(1, 1, 0), (1, -1, 0)$
8464	23	−1	1	1	6	$(1, -1, 0), (0, 1, 0)$
8649	−1	31	3	2	10	$(2, -1, 1), (2, 0, 1), (1, 0, -1), (1, 1, -1)$
9025	−19	5	1	2	24	$(1, 1, 0), (1, -1, 0)$

11.2.7   Some more quartic fields

The following tables contain data for quartic fields with certain signatures and Galois groups computed by the algorithm of I. Gaál, A. Pethő and M. Pohst [GPP93].

Our tables contain the following data. In the first column we list the discriminant D_K of the field $K = \mathbb{Q}(\xi)$. The second column contains the coefficients (a_1, a_2, a_3, a_4) of the minimal polynomial $f(x) = x^4 + a_1 x^3 + a_2 x^2 + a_3 x + a_4$ of ξ. In the third column we list the minimal m for which the index form equation $I(x_2, x_3, x_4) = \pm m$ has solutions with $|x_2|, |x_3|, |x_4| < 10^{10}$. It is followed by an integral basis of K in case the integral basis is not the power basis $\{1, \xi, \xi^2, \xi^3\}$. Finally, we list the solutions (x_2, x_3, x_4) with absolute value $< 10^{10}$ of the index form equation corresponding to the given integral basis and with right-hand side $\pm m$. Note that these data are complete with high probability, and m is certainly the minimal index of K. For totally complex fields the method of Section 6.4 was used which gives complete solutions. The fields are ordered according to the absolute value of the discriminant. We especially treat fields with Galois group A_4 and S_4 which count as more difficult from a computational point of view.

Totally real quartic fields with Galois group A_4

D_K	f	m	(integral basis) solutions
26569	$(-2, -7, 3, 8)$	1	$(-3, 3, 2), (-2, 1, 0), (0, 1, 0), (1, 0, 0), (3, 3, -1),$
			$(3, 7, -2), (4, 3, -1), (119, 114, -36)$
33489	$(0, -7, -3, 1)$	1	$(-7, 0, 1), (1, 0, 0)$
61504	$(-2, -7, 6, 11)$	1	$(1, 0, 0), (4, 3, -1), (68, 3, -9)$
76729	$(-1, -16, 3, 1)$	4	$1, \xi, \xi^2, (1 + 2\xi + 2\xi^2 + \xi^3)/4,$
			$(-426, 225, 625), (1, 0, 0), (2, 0, -1), (3, 1, -1), (4, 1, -1),$
			$(5, 1, -1), (14, 2, -3), (18, 3, -4), (22, 4, -5), (41, 6, -9),$
			$(129, 23, -29)$
121801	$(-1, -10, 3, 20)$	1	$(-3, 2, 1), (-1, 1, 0), (0, 1, 0), (1, 0, 0),$
			$(3, 3, -1)$
157609	$(0, -13, -2, 19)$	4	$1, \xi, (1 + \xi + \xi^2)/2, (1 + \xi^3)/2,$
			$(-6, 1, 1), (-1, 1, 0), (0, 1, 0), (0, 3, -1), (1, 0, 0),$
			$(5, 1, -1), (24, 7, -5)$
165649	$(-2, -9, -1, 3)$	1	$(1, 0, 0), (1, 4, -1), (8, 2, -1), (39, 6, -4)$
223729	$(-1, -16, -7, 27)$	2	$1, \xi, \xi^2, (1 + \xi^3)/2,$
			$(-8, 0, 1), (1, 0, 0), (10, 3, -2)$
261121	$(-1, -9, 2, 11)$	1	$(1, 0, 0)$
270400	$(-2, -10, 6, 19)$	1	$(1, 0, 0), (2, 4, -1)$
299209	$(-2, -19, 11, 10)$	1	$1, \xi, (-1 - \xi + \xi^2)/3, (-1 + \xi + \xi^3)/3,$
			$(0, 1, 0), (1, 6, -1), (7, 3, -1)$
346921	$(-1, -22, -16, 8)$	1	$1, \xi, (\xi + \xi^2)/2, (\xi^2 + \xi^3)/4,$
			$(5, 1, -1)$
368449	$(-2, -13, 21, 26)$	1	$(-11, 1, 1), (-10, 1, 1), (-9, 1, 1), (-2, 1, 0), (0, 1, 0),$
			$(1, 0, 0)$
373321	$(-1, -17, 38, -13)$	1	$(1, 0, 0)$
408321	$(-1, -12, 15, 12)$	1	$(-7, 2, 1), (1, 0, 0)$
423801	$(0, -13, -15, 4)$	1	$(1, 0, 0), (2599, 1020, -400)$
461041	$(-1, -28, 65, -25)$	4	$1, \xi, \xi^2, (5 + 2\xi + 4\xi^2 + \xi^3)/10,$
			$(-2, 0, 1), (8, 1, -3), (9, 1, -3), (27, 4, -9)$
473344	$(0, -10, -4, 6)$	1	$(1, 0, 0), (9, 1, -1)$
494209	$(-2, -19, 19, 19)$	9	$1, \xi, \xi^2, (2 - 2\xi + 2\xi^2 + \xi^3)/9,$
			$(-3, 1, 10), (-1, 0, 1), (-1, 1, 0), (1, -1, 1), (1, 0, 0),$
			$(3, 1, -2), (6, 1, -2), (7, 2, -5), (7, 2, -4), (8, 1, -4),$
			$(16, 4, -9), (18, 5, -10), (51, 7, -26)$
502681	$(0, -17, -13, 35)$	1	$(1, 0, 0), (9, 3, -1), (21, 5, -2)$
505521	$(-1, -24, 1, 11)$	10	$1, \xi, \xi^2, (-3 - 2\xi - 4\xi^2 + \xi^3)/10,$
			$(-1, 0, 1), (1, 0, 0)$
529984	$(-2, -22, 10, -1)$	1	$(1, 0, 0), (22, 2, -1)$
549081	$(-1, -13, 10, 4)$	2	$1, \xi, \xi^2, (\xi + \xi^2 + \xi^3)/2,$
			$(1, 0, 0), (7, 1, -1), (14, 1, -2)$
582169	$(-2, -29, -11, -1)$	1	$(1, 0, 0), (28, 2, -1), (29, 2, -1), (30, 2, -1), (114, 9, -4),$
			$(201, 15, -7)$
660969	$(-1, -16, 13, -2)$	1	$(1, 0, 0)$
727609	$(-1, -28, 31, -2)$	9	$1, \xi, (-1 + \xi + \xi^2)/3, (1 + \xi + \xi^3)/3,$
			$(-29, 0, 3), (-10, 0, 1), (-4, 4, 1), (-2, 1, 0), (1, 0, 0),$
			$(1, 1, 0), (9, 1, -1)$
848241	$(-1, -16, -11, 7)$	2	$1, \xi, \xi^2, (1 + \xi^3)/2,$
			$(1, 0, 0), (2, 2, -1), (7, 1, -1), (406, 15, -50)$
876096	$(0, -18, -8, 12)$	4	$1, \xi, \xi^2/2, \xi^3/2,$
			$(0, 1, 0), (1, 0, 0)$
877969	$(-1, -22, 39, 28)$	1	$(-16, 2, 1), (-15, 2, 1), (1, 0, 0)$
900601	$(-1, -19, 20, 44)$	4	$1, \xi, \xi^2, (2 - \xi + \xi^2 + \xi^3)/4,$
			$(-20, 5, 11), (-4, 0, 1), (-3, 0, 1), (1, 0, 0), (4, 1, -1)$

Totally real quartic fields with Galois group S_4

D_K	f	m	*(integral basis)* *solutions*
1957	$(0, -4, -1, 1)$	1	$(-12, 1, 3), (-8, 1, 2), (-5, 0, 1), (-4, 0, 1), (-4, 1, 1),$
			$(-3, 0, 1), (0, 1, 0), (0, 2, 1), (1, 0, 0), (1, 2, -1),$
			$(2, 1, -1), (3, 1, -1), (4, 1, -1), (4, 9, -5), (4, 33, 16),$
			$(8, 1, -2), (14, 3, -4)$
2777	$(-1, -4, 1, 2)$	1	$(-6, 2, 1), (-4, 0, 1), (-1, 1, 0), (-1, 1, 1), (0, -4, 1),$
			$(0, 1, 0), (1, 0, 0), (1, 2, -1), (1, 10, -4), (2, 2, -1),$
			$(3, 1, -1), (3, 2, -1), (5, 1, -1), (6, 3, -2), (21, 1, -5)$
3981	$(-1, -4, 2, 1)$	1	$(-4, 0, 1), (-2, 1, 0), (-1, 2, 0), (0, -1, 1), (1, 0, 0),$
			$(1, 1, 0), (3, 1, -1), (4, 1, -1), (15, 4, -4), (21, 1, -5)$
5744	$(0, -5, -2, 1)$	1	$(-5, 0, 1), (1, 0, 0), (1, 2, -1), (4, 1, -1), (71, 12, -16)$
6224	$(-2, -4, 2, 2)$	1	$(-1, 1, 1), (1, 0, 0), (1, 3, -1), (3, 2, -1), (5, 1, -1),$
			$(5, 5, -2), (7, 4, -2)$
6809	$(0, -5, -1, 1)$	1	$(-9, 1, 2), (-6, 0, 1), (-5, 0, 1), (-5, 1, 1), (-4, 0, 1),$
			$(0, 1, 0), (1, 0, 0), (1, 2, -1), (9, 1, -2)$
7053	$(-2, -4, 3, 3)$	1	$(1, 0, 0), (1, 3, -1), (2, -4, 1), (2, 2, -1), (5, 1, -1), (20, 24, -9)$
7537	$(-1, -5, 4, 3)$	1	$(-15, 7, 6), (-9, 1, 2), (-4, 0, 1), (-3, 1, 1), (-1, 1, 0),$
			$(0, 1, 0), (1, -3, 1), (1, 0, 0), (4, 1, -1), (6, 1, -1), (17, 6, -4)$
8069	$(-1, -5, 5, 1)$	1	$(-125, 56, 49), (-19, 1, 4), (-5, 0, 1), (-4, 0, 1), (-3, 1, 1),$
			$(-2, 1, 0), (-1, 1, 0), (0, 1, 0), (1, 0, 0), (1, 1, 0),$
			$(2, 2, -1), (5, 1, -1), (24, 6, -5)$
8468	$(-1, -5, 3, 4)$	1	$(-4, 0, 1), (1, 0, 0), (4, 2, -1)$
8789	$(-1, -6, -2, 1)$	1	$(-1, 1, 1), (1, 0, 0), (4, 2, -1), (6, 1, -1), (7, 1, -1),$
			$(9, 3, -2), (11, 3, -2), (14, 24, -9), (25, 3, -4), (161, 19, -26)$
9301	$(-1, -5, 1, 3)$	1	$(-5, 0, 1), (-1, 1, 0), (0, 1, 0), (1, 0, 0), (3, 1, -1),$
			$(3, 2, -1), (3, 16, -6), (234, 116, -65)$
9909	$(0, -6, -3, 3)$	1	$(-23, 2, 4), (-5, 0, 1), (1, 0, 0), (1, 3, 1), (2, 2, -1),$
			$(4, 1, -1), (4, 2, -1), (5, 1, -1), (5, 4, -2)$
10273	$(-2, -5, 1, 2)$	1	$(-2, 1, 0), (0, -7, 2), (0, 1, 0), (1, 0, 0), (2, 3, -1),$
			$(4, 2, -1), (10, 8, -3)$
10889	$(-1, -5, 2, 1)$	1	$(-6, 0, 1), (1, -3, 1), (1, 0, 0), (4, 1, -1), (5, 1, -1),$
			$(14, 4, -3), (47, 3, -9)$
11197	$(-2, -4, 3, 1)$	1	$(-1, 1, 1), (0, 3, -1), (1, 0, 0), (3, 2, -1), (4, 2, -1),$
			$(5, 1, -1)$
11324	$(-1, -5, 4, 2)$	1	$(-5, 0, 1), (1, 0, 0)$
11344	$(-2, -4, 4, 3)$	1	$(-2, 0, 1), (-2, 1, 1), (-1, 1, 0), (1, 0, 0), (2, 2, -1),$
			$(2, 3 - 1)$
11348	$(-1, -5, 1, 2)$	1	$(1, 0, 0)$
12197	$(-1, -5, 3, 1)$	1	$(-5, 0, 1), (1, -3, 1), (1, 0, 0), (5, 1, -1)$
12357	$(-1, -5, 3, 3)$	1	$(-5, 0, 1), (1, -3, 1), (1, 0, 0)$
13676	$(-1, -6, 7, 1)$	1	$(-9, 0, 2), (-4, 1, 1), (1, 0, 0), (6, 1, -1)$
13768	$(-1, -5, 2, 2)$	1	$(-5, 0, 1), (1, 0, 0), (17, 6, -4)$
14013	$(-1, -6, 6, 3)$	1	$(-4, 1, 1), (1, 0, 0)$
14197	$(0, -6, -3, 1)$	1	$(-6, 0, 1), (1, 0, 0), (2, 2, -1), (5, 1, -1)$
14272	$(-2, -5, 2, 3)$	1	$(1, 0, 0), (2, 3, -1), (6, 1, -1)$
14656	$(-2, -4, 4, 2)$	1	$(-1, 1, 0), (1, 0, 0), (3, -7, 2), (3, 2, -1), (5, 1, -1)$
15188	$(-1, -7, 1, 2)$	1	$1, \xi, \xi^2, (\xi + \xi^3)/2$
			$(1, -2, 1), (2, 1, -1), (4, 1, -1), (12, 1, -3)$
15529	$(-1, -6, -1, 2)$	1	$(0, 2, 1), (1, 0, 0), (4, 2, -1), (25, 2, -4)$
15952	$(0, -6, -2, 1)$	1	$(-47, 2, 8), (-6, 0, 1), (1, 0, 0)$

Quartic fields of mixed signature

D_K	f	m	(integral basis) solutions
-275	$(-1, 0, 2, -1)$	1	$(0, 0, 1), (0, 1, -1), (1, 0, 0), (1, 2, -4), (2, -2, 1)$
-283	$(0, 0, -1, -1)$	1	$(-1, 0, 1), (-1, 2, 0), (0, -2, 1), (0, -1, 1), (0, 0, 1),$
			$(0, 1, 0), (1, -1, 1), (1, 0, 0,), (1, 0, 1), (1, 1, 1),$
			$(1, 2, 0), (2, -3, 4), (6, 5, 4)$
-331	$(-1, -1, 1, -1)$	1	$(-2, 0, 1), (-2, 3, 0), (-1, 0, 1), (-1, 1, 0), (-1, 2, 4),$
			$(-1, 3, 0), (0, 0, 1), (0, 1, 0), (1, -2, 1), (1, 0, 0), (1, 1, -1)$
-400	$(0, -1, 0, -1)$	1	$(-1, 0, 1), (0, 1, -1), (0, 1, 1), (1, 0, 0)$
-448	$(-2, 1, 2, -1)$	1	$(0, 1, -1), (1, -6, 4), (1, -2, 1), (1, 0, 0), (4, -3, 1)$
-475	$(-1, -2, -2, -1)$	1	$(0, 2, -1), (1, 0, 0), (2, 1, -1)$
-491	$(-1, -2, 2, -1)$	1	$(-35, 29, 42), (-2, 0, 1), (-1, 1, 0), (-1, 1, 1), (0, 1, 0),$
			$(1, 0, 0), (2, 1, -1), (5, 1, -2), (7, -10, 4)$
-507	$(-1, -1, -1, 1)$	1	$(-1, 0, 1), (1, 0, 0), (1, 1, -1)$
-563	$(-1, -1, -1, -1)$	1	$(-1, 1, 0), (0, -2, 1), (0, 1, 0), (1, -2, 1), (1, 0, 0),$
			$(1, 1, -1), (1, 1, 1), (15, -71, 40)$
-643	$(-1, 0, -2, 1)$	1	$(-2, 1, 0), (0, -1, 1), (0, 0, 1), (1, 0, 0), (1, 1, -1),$
			$(1, 1, 0), (1, 2, -4), (14, 9, 16)$
-688	$(0, 0, -2, -1)$	1	$(-1, 1, 0), (0, 0, 1), (0, 1, -1), (1, -2, 4), (1, 0, 0),$
			$(1, 1, 0), (9, -19, 40)$
-731	$(0, -2, -1, -1)$	1	$(-2, 0, 1), (0, -1, 1), (0, 1, 0), (1, 0, 0), (1, 1, -1),$
			$(1, 2, 1), (2, 1, -1)$
-751	$(-1, -1, -2, -1)$	1	$(0, -1, 1), (0, 2, -1), (1, 0, 0), (1, 1, -1), (1, 1, 1),$
			$(1, 6, -4), (2, 1, -1)$
-775	$(-1, 0, -3, -1)$	2	$1, \xi, \xi^2, (1 + \xi^3)/2$
			$(0, -1, 2), (0, 0, 1), (1, -1, 1), (1, 0, 0), (2, -6, 9), (2, 1, 2)$
-848	$(0, -1, -2, 1)$	1	$(-1, 0, 1), (0, 1, 1), (1, 0, 0)$
-976	$(0, -3, -2, -1)$	1	$(-3, 0, 1), (1, 0, 0), (1, 2, 1), (2, 1, -1)$
-1024	$(0, -2, 0, -1)$	1	$(-2, 0, 1), (1, 0, 0)$
-1099	$(-1, 1, -3, 1)$	1	$(0, -3, 2), (1, -1, 1), (1, 0, 0), (1, 0, 1)$
-1107	$(-1, 0, -2, -1)$	1	$(0, -2, 1), (0, -1, 1), (1, 0, 0), (2, -4, 3), (3, 2, 2)$
-1156	$(-1, -2, -1, 1)$	1	$(0, 1, 1), (1, 0, 0), (2, 1, -1), (9, 2, -4)$
-1192	$(-1, -2, 1, -1)$	1	$(-1, 2, 0), (0, 1, 1), (1, 0, 0), (2, 1, -1)$
-1255	$(0, -1, -3, -1)$	1	$(-1, 0, 1), (1, 0, 0), (1, 1, -1), (1, 1, 1), (2, 1, -2)$
-1323	$(-1, -3, -1, 1)$	1	$(1, 0, 0), (1, 2, -1), (3, 1, -1)$
-1328	$(0, -3, -2, 1)$	1	$(-26, 3, 9), (-3, 0, 1), (1, 0, 0), (1, 2, 1), (2, 1, -1)$
-1371	$(0, 2, -1, -1)$	1	$(0, 1, 0), (1, 0, 0), (2, -1, 1), (2, 0, 1), (3, 1, 1),$
			$(9, -2, 4)$
-1375	$(-2, -1, 2, -4)$	4	$1, \xi, (\xi + \xi^2)/2, (\xi + \xi^3)/2$
			$(-6, 3, 8), (-2, 1, 0), (-1, 0, 1), (0, 1, 0), (0, 2, -1),$
			$(1, 0, 0), (1, 1, -1), (2, -3, 1), (24, -27, 8)$
-1399	$(-1, 0, 1, -2)$	1	$(-1, 1, 0), (-1, 2, 1), (0, 0, 1), (0, 1, 0), (1, 0, 0),$
			$(1, 3, -1), (2, -2, 1)$
-1423	$(-1, 1, -2, -1)$	1	$(-1, 1, 0), (0, 1, 0), (1, -1, 1), (1, 0, 0), (3, -3, 2),$
			$(7, 2, 4)$
-1424	$(0, 1, -2, -1)$	1	$(1, 0, 0), (1, 0, 1), (2, 1, 1)$
-1456	$(0, -2, -2, 1)$	1	$(-2, 0, 1), (1, 0, 0)$

Totally complex quartic fields with Galois group A_4

D_K	f	m	(integral basis) solutions
3136	$(-2, 2, 0, 2)$	1	$(1, -1, 0), (1, 0, 0)$
4225	$(-2, 3, 3, 1)$	1	$(1, 0, 0), (3, -2, 1), (4, -2, 1)$
5184	$(0, 6, -8, 9)$	1	$1, \xi, (1 + \xi^2)/2, (-1 - \xi + \xi^2 + \xi^3)/4$
			$(1, -1, 1), (2, 0, 1)$
8281	$(-1, 5, -4, 3)$	1	$(0, 1, 0), (1, -1, 0), (1, 0, 0)$
10816	$(-2, 2, 4, 2)$	1	$(1, 0, 0)$
15376	$(0, 7, -2, 14)$	2	$1, \xi, \xi^2, (\xi^2 + \xi^3)/2$
			$(1, 0, 0)$
17689	$(0, 3, -7, 4)$	1	$(1, 0, 0), (3, 1, 1), (4, 1, 0)$
23104	$(0, 10, -8, 17)$	1	$1, \xi, (1 + \xi^2)/2, (1 + \xi + \xi^2 + \xi^3)/4$
			$(1, -2, 1), (2, 0, 1)$
23409	$(-1, 6, -5, 8)$	1	$(1, 0, 0), (4, -2, 1)$
29241	$(-1, 9, -2, 23)$	3	$1, \xi, \xi^2, (-1 - \xi + \xi^2 + \xi^3)/3$
			$(1, -1, 1), (1, 0, 0), (1, 0, 1)$
29584	$(-2, 2, 10, 9)$	4	$1, \xi, (1 + \xi^2)/2, (\xi + \xi^3)/2$
			$(1, 0, 0), (2, -3, 1), (2, 1, -1)$
40401	$(-1, 4, -12, 24)$	3	$1, \xi, (\xi + \xi^2)/2, (2\xi + \xi^2 + \xi^3)/4$
			$(0, 1, -1), (2, -1, 0)$
41209	$(-1, 5, -6, 7)$	1	$(1, 0, 0), (4, 0, 1)$
43681	$(-1, -9, 6, 25)$	1	$(1, 0, 0), (3, 3, -1)$
47961	$(-1, -2, 0, 24)$	3	$1, \xi, (\xi + \xi^2)/2, (\xi^2 + \xi^3)/4$
			$(1, -1, 0), (1, 1, -1), (1, 1, 0)$
61504	$(-2, 11, -2, 9)$	4	$1, \xi, (1 + \xi + \xi^2)/2, (1 + \xi^3)/2$
			$(1, -1, 0), (1, 0, 0), (6, -2, 1)$
63504	$(-2, 9, 0, 9)$	3	$1, \xi, \xi^2, (\xi^2 + \xi^3)/3$
			$(1, 0, 0), (3, -1, 1)$
74529	$(-1, 7, 9, 24)$	8	$1, \xi, (\xi + \xi^2)/2, (-\xi + \xi^3)/4$
			$(1, 0, 0)$
82369	$(-1, 4, -7, 8)$	1	$(1, 0, 0)$
87616	$(0, 14, -8, 33)$	8	$1, \xi, (1 + \xi^2)/2, (-1 - \xi + \xi^2 + \xi^3)/4$
			$(0, 1, 0), (1, 0, 0)$
90601	$(-1, 11, 0, 43)$	7	$1, \xi, \xi^2, (2 + 3\xi + 2\xi^2 + \xi^3)/7$
			$(1, 0, 0)$
95481	$(0, 13, -3, 46)$	3	$1, \xi, \xi^2, (1 - \xi - \xi^2 + \xi^3)/3$
			$(1, 0, 0)$
99225	$(-1, 0, 7, 14)$	3	$1, \xi, \xi^2, (-1 - \xi + \xi^2 + \xi^3)/3$
			$(1, -1, 1), (1, 0, 0)$
110889	$(-1, 6, 11, 10)$	3	$1, \xi, \xi^2, (-1 + \xi^3)/3$
			$(1, 0, 0)$
112225	$(0, 11, -5, 29)$	3	$1, \xi, \xi^2, (1 + \xi^2 + \xi^3)/3$
			$(1, 0, 0)$
118336	$(-2, 9, -6, 7)$	1	$(1, 0, 0)$
118336	$(0, 18, -32, 16)$	8	$1, \xi, \xi^2/2, (2\xi + \xi^3)/4$
			$(1, 0, 0), (4, 0, 1), (8, 1, 2)$
132496	$(-2, 14, -12, 20)$	4	$1, \xi, \xi^2/2, \xi^3/2$
			$(1, -2, 0), (1, 0, 0), (5, 0, 1)$
159201	$(-2, 7, 21, 63)$	8	$1, \xi, (-\xi + \xi^2)/3, (3\xi - \xi^2 + \xi^3)/9$
			$(1, 0, -1)$
162409	$(-1, -6, 7, 18)$	1	$(1, 0, 0)$

Totally complex quartic fields with Galois group S_4

D_K	f	m	(integral basis) solutions
229	$(0, 0, -1, 1)$	1	$(0, 0, 1)$, $(0, 1, -1)$, $(0, 1, 0)$, $(0, 1, 1)$, $(0, 1, 2)$, $(0, 2, 1)$, $(1, 0, -1)$, $(1, 0, 0)$, $(1, 0, 1)$, $(1, 1, 1)$
257	$(0, 1, -1, 1)$	1	$(0, 0, 1)$, $(0, 1, -2)$, $(0, 1, 0)$, $(1, -1, 0)$, $(1, 0, 0)$, $(1, 0, 1)$, $(1, 1, 0)$, $(1, 1, 1)$, $(2, 0, 1)$, $(4, 3, 2)$
592	$(0, 2, -2, 1)$	1	$(1, -1, 0)$, $(1, 0, 0)$, $(1, 1, 0)$, $(2, 0, 1)$, $(2, 1, 1)$
697	$(-1, 2, -1, 2)$	1	$(0, 1, -2)$, $(0, 1, 0)$, $(1, -2, 0)$, $(1, -1, 0)$, $(1, -1, 1)$, $(1, 0, 0)$, $(1, 0, 1)$
761	$(-1, 1, 2, 1)$	1	$(1, -1, 1)$, $(1, 0, 0)$, $(1, 1, 0)$, $(2, -2, 1)$, $(2, -1, 0)$, $(2, -1, 1)$
788	$(-1, 2, -2, 2)$	1	$(1, -1, 1)$, $(1, 0, 0)$
892	$(-1, -1, 0, 2)$	1	$(1, 0, -1)$, $(1, 0, 0)$
985	$(-1, 2, -3, 2)$	1	$(1, -1, 1)$, $(1, 0, 0)$, $(1, 0, 1)$, $(2, 0, 1)$
1016	$(-1, 1, -2, 2)$	1	$(1, 0, 0)$, $(1, 0, 1)$
1076	$(-1, 3, -3, 2)$	1	$(1, 0, 0)$, $(2, -2, 1)$, $(2, 0, 1)$
1129	$(-1, 0, -1, 2)$	1	$(0, 0, 1)$, $(1, 0, 0)$
1229	$(-1, 3, -1, 3)$	1	$(0, 1, 0)$, $(1, -1, 0)$, $(1, 0, 0)$, $(2, -1, 2)$, $(2, 2, 1)$
1257	$(0, -1, -1, 2)$	1	$(0, 1, 0)$, $(0, 1, 1)$, $(1, 0, 0)$
1264	$(0, 3, -2, 1)$	1	$(1, 0, 0)$, $(2, -1, 1)$, $(3, 0, 1)$
1384	$(-1, -1, 2, 2)$	1	$(1, -2, 1)$, $(1, 0, 0)$
1396	$(-1, 1, -1, 2)$	1	$(1, 0, 0)$
1436	$(-1, 3, 0, 2)$	1	$(1, 0, 0)$, $(1, 0, 1)$
1489	$(-1, 4, -1, 2)$	1	$(0, 1, 0)$, $(1, -1, 0)$, $(1, 0, 0)$, $(3, -1, 1)$
1492	$(-1, 2, 0, 2)$	1	$(1, -1, 1)$, $(1, 0, 0)$
1509	$(-1, 2, 2, 1)$	1	$(1, 0, 0)$, $(2, -1, 1)$, $(3, -1, 1)$
1556	$(-1, -1, 1, 2)$	1	$(1, 0, 0)$
1593	$(-2, 3, -1, 2)$	1	$(0, 1, -1)$, $(1, -1, 0)$, $(1, 0, 0)$
1616	$(0, 0, -2, 2)$	1	$(1, 0, -1)$, $(1, 0, 0)$, $(1, 1, 1)$
1765	$(-1, -1, -1, 3)$	1	$(0, 0, 1)$, $(1, 0, -1)$, $(1, 0, 0)$
1825	$(0, 1, -1, 2)$	1	$(0, 0, 1)$, $(0, 1, 0)$, $(1, 0, 0)$
1929	$(-1, 4, -3, 2)$	1	$(0, 1, 0)$, $(1, -1, 0)$, $(1, 0, 0)$
1937	$(0, 1, -3, 2)$	1	$(1, 0, 0)$, $(1, 1, 1)$, $(2, 1, 1)$
1940	$(-1, 0, 0, 2)$	1	$(1, -2, 0)$, $(1, 0, 0)$, $(1, 1, -1)$
2021	$(0, 0, -1, 2)$	1	$(0, 1, 0)$, $(1, 0, 0)$, $(4, 1, -2)$
2057	$(-1, 0, 3, 2)$	1	$(1, -2, 1)$, $(1, 0, 0)$, $(2, -2, 1)$

11.3   Sextic fields

11.3.1   *Totally real cyclic sextic fields*

Using the method of Section 8.1.2 we computed all generators of power integral bases in the five totally real cyclic sextic fields with smallest discriminants. The input data were taken from A.M. Bergé, J. Martinet and M. Olivier [BMO90].

In the following examples K is a totally real cyclic sextic number field with discriminant D_K and quadratic subfield M. An integral basis of K is $\{1, \vartheta, \vartheta^2, \omega, \omega\vartheta, \omega\vartheta^2\}$ where ω is explicitly given and ϑ is a root of the relative defining polynomial $f(t)$ over M. These are followed by the list of solutions of the index form equation (8.2).

It is clear from the tables of M. Pohst and H. Zassenhaus [PZ89] that the fields with discriminants 300125, 371293, 453789 and 1075648 admit power integral bases. In case of the field with discriminant 820125 the generating element given in [PZ89] has index > 1, but also in this case we found several solutions of the index form equation, that is, elements with index 1.

$D_K = 300125$, $M = \mathbb{Q}(\sqrt{5})$, $\omega = (1 + \sqrt{5})/2$, $f(t) = t^3 - (7 + 7\omega)t + (7 + 14\omega)$
Solutions: $(x_1, x_2, y_0, y_1, y_2) = (-71, 68, 66, 44, -42)$, $(-61, 73, 88, 38, -45)$,
$(-12, 11, 13, 7, -7)$, $(-11, 13, 15, 7, -8)$, $(-10, -5, 4, 6, 3)$, $(-6, 6, 9, 3, -4)$,
$(-6, 6, 10, 3, -4)$, $(-5, 4, 9, 2, -3)$, $(-5, 5, 5, 3, -3)$, $(-4, 3, 4, 2, -2)$,
$(-4, 3, 5, 2, -2)$, $(-4, 4, 9, 1, -3)$, $(-4, 5, 5, 3, -3)$, $(-3, 2, 9, 0, -2)$,
$(-3, 2, 10, 0, -2)$, $(-2, 1, 4, -2, 0)$, $(-2, 2, 1, 1, -1)$, $(-2, 3, 4, 1, -2)$,
$(-2, 3, 5, 1, -2)$, $(1, -1, -5, 0, 1)$, $(1, -1, -4, 0, 1)$, $(1, -1, 5, -2, 0)$,
$(1, 1, -15, 3, 1)$, $(1, 1, -5, 1, 0)$, $(1, 2, -1, 0, -1)$, $(2, -1, -5, 0, 1)$,
$(2, -1, -4, 0, 1)$, $(3, -1, -13, 2, 2)$, $(8, 5, -88, 15, 6)$, $(10, 4, -66, 17, 6)$.

$D_K = 371293$, $M = \mathbb{Q}(\sqrt{13})$, $\omega = (1 + \sqrt{13})/2$, $f(t) = t^3 - \omega t^2 + (-10 + 5\omega)t + (2 - \omega)$
Solutions: $(x_1, x_2, y_0, y_1, y_2) = (-499, 284, 121, -383, 218)$,
$(-456, 241, 136, -350, 185)$, $(-99, 56, 24, -76, 43)$, $(-82, 43, 25, -63, 33)$,
$(-46, 26, 11, -35, 20)$, $(-43, 43, 8, -33, 33)$, $(-42, 22, 12, -32, 17)$,
$(-31, 17, 9, -24, 13)$, $(-22, 13, 5, -17, 10)$, $(-17, 9, 4, -13, 7)$,
$(-17, 9, 5, -13, 7)$, $(-17, 13, 4, -13, 10)$, $(-16, 9, 3, -12, 7)$, $(-16, 9, 4, -12, 7)$,
$(-15, 8, 4, -11, 6)$, $(-14, 8, 4, -11, 6)$, $(-14, 8, 5, -11, 6)$, $(-11, 7, 5, -9, 5)$,
$(-9, 4, 3, -7, 3)$, $(-9, 5, 2, -7, 4)$, $(-9, 5, 2, -7, 4)$, $(-8, 4, 2, -6, 3)$,
$(-8, 4, 3, -6, 3)$, $(-7, 4, 2, -6, 3)$, $(-6, 4, 1, -5, 3)$, $(-6, 4, 2, -5, 3)$,
$(-4, 1, 1, -2, 1)$, $(-4, 4, 1, -3, 3)$, $(-1, 1, 1, -2, 1)$, $(-1, 2, 1, -3, 1)$,
$(0, 1, 5, -1, 0)$, $(1, 0, 0, 1, 0)$, $(4, 4, 24, -3, -1)$, $(6, -2, 0, -1, 0)$,
$(10, 2, 1, -4, -1)$.

$D_K = 453789$, $M = \mathbb{Q}(\sqrt{21})$, $\omega = (1 + \sqrt{21})/2$, $f(t) = t^3 - \omega t^2 + (-1 + \omega)t + (-3 + \omega)$
Solutions: $(x_1, x_2, y_0, y_1, y_2) = (-52, 25, 4, -29, 14)$, $(-43, 16, 13, -24, 9)$,
$(-12, 7, 2, -7, 3)$, $(-11, 5, 4, -6, 2)$, $(-9, 9, 1, -5, 5)$, $(-8, -3, 2, 3, 1)$,
$(-7, 3, 0, -4, 2)$, $(-5, 1, 1, -1, 1)$, $(-5, 2, 2, -3, 1)$, $(-5, 3, 2, -3, 1)$,
$(-4, 1, 0, -2, 1)$, $(-3, 14, 12, 1, -5)$, $(-2, 2, 2, -1, 0)$, $(-1, 2, 0, -1, 1)$,
$(-1, 4, 3, 0, -1)$, $(0, -1, 0, 1, 0)$, $(0, 1, 0, 0, 0)$, $(1, -2, -1, 1, 0)$,
$(1, 0, 0, 0, 0)$, $(1, 1, 1, 1, -1)$, $(1, 2, 3, 0, -1)$, $(2, -1, -1, 1, 0)$,
$(2, -1, 0, 1, -1)$, $(2, 1, 1, 1, -1)$, $(3, -2, 0, 2, -1)$, $(4, 3, 4, 1, -2)$,
$(5, 17, 20, -2, -6)$.

$D_K = 820125$, $M = \mathbb{Q}(\sqrt{5})$, $\omega = (1 + \sqrt{5})/2$, $f(t) = t^3 + (-6 - 6\omega)t + (6 + 11\omega)$
Solutions: $(x_1, x_2, y_0, y_1, y_2) = (-10, 8, 4, 6, -5)$, $(-4, 3, 2, 2, -2)$,
$(-4, 11, 16, 2, -7)$, $(-1, 0, 0, 1, 0)$, $(-1, 1, 3, 0, -1)$, $(-1, 1, 5, 0, -1)$,
$(0, -2, 8, -3, 0)$, $(0, 0, -2, 1, 0)$, $(0, 0, 8, -2, -1)$, $(1, 0, -4, 2, 0)$,
$(1, 1, -5, 1, 0)$, $(1, 1, -3, 1, 0)$, $(1, 2, -8, 2, 0)$, $(2, -4, -10, 0, 3)$,
$(2, -1, -12, 2, 2)$, $(2, -1, -4, 0, 1)$, $(2, 0, -9, 1, 1)$, $(2, 0, -7, 1, 1)$,
$(2, 1, 2, -2, -1)$, $(3, 2, -26, 5, 2)$, $(3, 2, -22, 6, 2)$, $(6, 3, 0, -4, -2)$,
$(8, 4, -68, 13, 6)$, $(9, 4, -60, 15, 6)$.

$D_K = 1075648$, $M = \mathbb{Q}(\sqrt{7})$, $\omega = \sqrt{7}$, $f(t) = t^3 - \omega t^2 + \omega$
Solutions: $(x_1, x_2, y_0, y_1, y_2) = (-6, -2, 1, 2, 1)$, $(-6, 2, 1, -2, 1)$,
$(-3, 0, -1, 0, 1)$, $(1, 0, 0, 0, 0)$, $(2, -4, 1, 1, -1)$, $(2, 0, 2, 0, -1)$,
$(2, 4, 1, -1, -1)$, $(3, -5, 2, 2, -1)$, $(3, 5, 2, -2, -1)$, $(5, -8, 4, 2, -3)$,
$(5, 8, 4, -2, -3)$, $(11, -10, 7, 4, -4)$, $(11, 10, 7, -4, -4)$.

11.3.2   Sextic fields with imaginary quadratic subfields

The following table lists data calculated by using the method of Section 8.1.3. We
give all generators of power integer bases of the 25 sextic fields K with imaginary
quadratic subfield M of smallest discriminants D_K (in absolute value). The input
data were taken from M. Olivier [Oli89]. We list the integral basis element ω of the
quadratic subfield M, the relative defining polynomial $f(t) \in \mathbb{Z}_M$ of the generator
element ϑ of K over M. In all these examples $\{1, \vartheta, \vartheta^2, \omega, \omega\vartheta, \omega\vartheta^2\}$ is an integral
basis of M and the solutions $(x_1, x_2, y_0, y_1, y_2)$ refer to the index form equation
(8.2).

$D_K = -9747$, $\omega = (1 + i\sqrt{3})/2$, $f(t) = t^3 + (-1 - \omega)t^2 + \omega t + (1 - \omega)$
Solutions: $(x_1, x_2, y_0, y_1, y_2) = (-2\,0\,1\,2\,-1)$, $(-1\,0\,0\,1\,-1)$, $(-1\,0\,-1\,2\,-1)$,
$(0\,0\,0\,1\,-1)$, $(0\,0\,-1\,2\,-1)$, $(-1\,0\,0\,0\,0)$, $(-1\,0\,1\,0\,0)$, $(-1\,0\,-1\,1\,0)$,
$(-1\,0\,0\,1\,0)$, $(0\,0\,0\,-1\,0)$, $(0\,0\,1\,-1\,0)$, $(-2\,1\,0\,1\,-1)$, $(-2\,1\,-2\,2\,-1)$,
$(-1\,1\,0\,0\,-1)$, $(-1\,1\,0\,1\,-1)$, $(0\,1\,0\,0\,-1)$, $(-1\,1\,1\,-1\,0)$, $(-1\,1\,0\,0\,0)$,
$(0\,1\,1\,-2\,0)$, $(0\,1\,0\,-1\,0)$, $(0\,1\,1\,0\,0)$.

$D_K = -10816$, $\omega = i$, $f(t) = t^3 + (-1 - \omega)t^2 + 5\omega t + (-1 - 4\omega)$
Solutions: $(x_1, x_2, y_0, y_1, y_2) = (-3\,0\,1\,0\,-2)$, $(-2\,0\,0\,0\,-1)$, $(-2\,0\,1\,0\,-1)$,
$(-1\,0\,0\,0\,-1)$, $(-1\,0\,1\,0\,-1)$, $(0\,0\,1\,-1\,0)$, $(2\,1\,2\,-1\,1)$, $(3\,1\,2\,-1\,2)$,
$(-3\,1\,4\,-2\,-2)$, $(-3\,1\,5\,-2\,-2)$, $(-1\,1\,3\,-2\,-1)$, $(-1\,1\,4\,-2\,-1)$, $(0\,1\,3\,-2\,0)$,
$(0\,1\,3\,-1\,0)$, $(-1\,2\,6\,-3\,-1)$, $(-1\,2\,7\,-3\,-1)$, $(0\,2\,6\,-3\,0)$, $(0\,2\,7\,-3\,0)$,
$(2\,2\,6\,-3\,1)$, $(5\,2\,4\,-3\,3)$, $(5\,2\,5\,-2\,3)$.

$D_K = -11691$, $\omega = (1 + i\sqrt{3})/2$, $f(t) = t^3 + (-1 - \omega)t^2 + (-2 + 2\omega)t + 1$
Solutions: $(x_1, x_2, y_0, y_1, y_2) = (-1\,0\,0\,1\,-1)$, $(-1\,0\,1\,1\,-1)$, $(-1\,0\,0\,2\,-1)$,
$(0\,0\,0\,1\,-1)$, $(0\,0\,-1\,2\,-1)$, $(-1\,0\,0\,0\,0)$, $(-1\,0\,-1\,1\,0)$, $(-1\,0\,0\,1\,0)$,
$(-2\,1\,1\,1\,-1)$, $(-2\,1\,1\,2\,-1)$, $(-1\,1\,1\,0\,-1)$, $(-1\,1\,2\,0\,-1)$, $(-1\,1\,1\,1\,-1)$,
$(-1\,1\,1\,-1\,0)$, $(-1\,1\,2\,-1\,0)$, $(-1\,1\,0\,0\,0)$, $(-1\,1\,1\,0\,0)$, $(1\,1\,0\,-1\,0)$,
$(2\,-2\,-3\,0\,1)$, $(0\,1\,1\,-2\,1)$.

$D_K = -12167$, $\omega = (1 + i\sqrt{23})/2$, $f(t) = t^3 + (-1 - \omega)t^2 + (-2 + \omega)t + 1$
Solutions: $(x_1, x_2, y_0, y_1, y_2) = (-1\,0\,0\,0\,0)$, $(-1\,1\,1\,-1\,0)$, $(0\,1\,0\,-1\,0)$.

$D_K = -14283, \omega = (1 + i\sqrt{3})/2, f(t) = t^3 + (1 - \omega)t - 1$
Solutions: $(x_1, x_2, y_0, y_1, y_2) = (0\ 0\ -1\ -1\ -1), (1\ 0\ 1\ 0\ -1), (-1\ 0\ 0\ 0\ 0),$
$(-1\ 0\ 1\ 0\ 0), (0\ 0\ 0\ -1\ 0), (0\ 0\ 1\ -1\ 0), (0\ 1\ -1\ 0\ -1), (0\ 1\ -1\ 1\ -1),$
$(1\ 1\ 0\ 0\ -1), (0\ 1\ -1\ 0\ 0), (0\ 1\ 0\ 1\ 0), (1\ 1\ -1\ 0\ 0), (-2\ 3\ -4\ 3\ 0),$
$(-3\ -3\ 4\ 2\ 3).$

$D_K = -16551, \omega = (1 + i\sqrt{3})/2, f(t) = t^3 + (-1 - \omega)t^2 + 2t + (-1 + \omega)$
Solutions: $(x_1, x_2, y_0, y_1, y_2) = (-3\ 0\ -1\ 3\ -2), (-2\ 0\ -1\ 2\ -1), (-1\ 0\ -1\ 1\ -1),$
$(-1\ 0\ -1\ 1\ 0), (0\ 0\ 0\ -1\ 0), (-2\ 1\ 0\ 0\ -1), (-2\ 1\ -1\ 1\ -1), (-1\ 1\ -1\ 0\ -1),$
$(-1\ 1\ 0\ -1\ 0), (0\ 1\ 0\ -2\ 0), (1\ 1\ 3\ -4\ 1).$

$D_K = -16807, \omega = (1 + i\sqrt{7})/2, f(t) = t^3 - \omega t^2 + (-1 + \omega)t + 1$
Solutions: $(x_1, x_2, y_0, y_1, y_2) = (-1\ 0\ 0\ 1\ -1), (-1\ 0\ 0\ 0\ 0), (-2\ 1\ 0\ 1\ -1),$
$(-1\ 1\ 0\ 0\ -1), (-1\ 1\ 1\ 0\ 0), (0\ 1\ 1\ -1\ 0), (0\ 1\ 0\ 0\ 0), (1\ 1\ 0\ -1\ 0),$
$(-1\ 2\ 1\ 0\ -1).$

$D_K = -19683, \omega = (1 + i\sqrt{3})/2, f(t) = t^3 + (-1 + \omega)$
Solutions: $(x_1, x_2, y_0, y_1, y_2) = (0\ 0\ 0\ 0\ -1), (1\ 0\ 1\ 0\ -1), (-1\ 0\ 0\ 0\ 0),$
$(-1\ 0\ 0\ 1\ 0), (0\ 0\ 0\ -1\ 0), (0\ 1\ 0\ 0\ -1), (1\ 1\ -1\ -1\ -1), (0\ 1\ 1\ -1\ 0),$
$(0\ 1\ 0\ 0\ 0).$

$D_K = -21168, \omega = (1 + i\sqrt{3})/2, f(t) = t^3 - t^2 + (1 - 2\omega)t + 1$
Solutions: $(x_1, x_2, y_0, y_1, y_2) = (-1\ 0\ 1\ 1\ -1), (-1\ 0\ 0\ 2\ -1), (0\ 0\ 1\ 1\ -1),$
$(-1\ 0\ 0\ 1\ 0), (0\ 0\ 0\ -1\ 0), (-2\ 1\ -1\ 1\ -1), (-1\ 1\ 0\ 0\ -1), (-1\ 1\ -1\ 1\ -1),$
$(-1\ 1\ -1\ 0\ 0).$

$D_K = -21296, \omega = (1 + i\sqrt{11})/2, f(t) = t^3 - \omega t^2 + (-1 + \omega)t + 1$
Solutions: $(x_1, x_2, y_0, y_1, y_2) = (-2\ 0\ 0\ 1\ -1), (-1\ 0\ -1\ 1\ -1), (-1\ 0\ 0\ 0\ 0),$
$(-1\ 1\ 1\ 0\ 0), (0\ 1\ 1\ -1\ 0), (1\ 1\ 0\ -1\ 0), (0\ 1\ 1\ -1\ 1), (-1\ 2\ 1\ -1\ 0),$
$(1\ 2\ 2\ -2\ 1).$

$D_K = -22592, \omega = i, f(t) = t^3 + (-1 - \omega)t^2 + (1 + 2\omega)t - \omega$
Solutions: $(x_1, x_2, y_0, y_1, y_2) = (-1\ 0\ 0\ 0\ -1), (-1\ 0\ -1\ 1\ -1), (-1\ 0\ 0\ 1\ -1),$
$(-1\ 0\ 0\ 0\ 0), (-1\ 1\ 1\ -1\ 0), (-1\ 1\ 0\ 0\ 0), (0\ 1\ 1\ -1\ 0), (0\ -1\ -2\ 1\ -1),$
$(-1\ 1\ 0\ 0\ -1), (-1\ -2\ -6\ 3\ -3).$

$D_K = -22707, \omega = (1 + i\sqrt{3})/2, f(t) = t^3 + (-1 - \omega)t^2 + 2\omega t + (1 - 2\omega)$
Solutions: $(x_1, x_2, y_0, y_1, y_2) = (-1\ 0\ -1\ 1\ -1), (-1\ 0\ 0\ 0\ 0), (-1\ 0\ 0\ 1\ 0),$
$(-1\ 1\ 1\ 0\ 0).$

$D_K = -23031, \omega = (1 + i\sqrt{3})/2, f(t) = t^3 - t^2 + (-1 + \omega)$
Solutions: $(x_1, x_2, y_0, y_1, y_2) = (-1\ 0\ -1\ 2\ -1), (0\ 0\ 0\ 0\ -1), (0\ 0\ 0\ 1\ -1),$
$(1\ 0\ 1\ 0\ -1), (-1\ 0\ 0\ 0\ 0), (-1\ 0\ 0\ 1\ 0), (-1\ 1\ 0\ 1\ -1), (0\ 1\ 1\ 0\ -1).$

$D_K = -24003, \omega = (1 + i\sqrt{3})/2, f(t) = t^3 - t^2 - t + (1 - \omega)$
Solutions: $(x_1, x_2, y_0, y_1, y_2) = (0\ 0\ 1\ 0\ -1), (0\ 0\ 1\ 1\ -1), (-1\ 0\ 0\ 0\ 0),$
$(-1\ 0\ 0\ 1\ 0), (0\ 0\ 0\ -1\ 0), (0\ 0\ 1\ -1\ 0), (-2\ 1\ -1\ 2\ -1), (-1\ 1\ 1\ 1\ -1),$
$(0\ 1\ 1\ 0\ -1), (-1\ 1\ 0\ 0\ 0), (0\ 1\ 0\ -1\ 0).$

$D_K = -25947, \omega = (1 + i\sqrt{3})/2, f(t) = t^3 + t + 1$
(Remark: in this case a generator of K is $\vartheta\omega$ where ϑ is a root of f)
Solutions: $(x_1, x_2, y_0, y_1, y_2) = (0\ 0\ -1\ 0\ -1), (-1\ 0\ 0\ 1\ 0), (0\ 0\ 0\ -1\ 0),$
$(0\ 1\ -1\ 0\ -1), (0\ 1\ 1\ 1\ 0), (1\ 1\ -1\ -1\ 0).$

$D_K = -29791, \omega = (1 + i\sqrt{31})/2,$
$f(t) = t^3 + (-1 - \omega)t^2 + (-2 + \omega)t + 1$
Solutions: $(x_1, x_2, y_0, y_1, y_2) = (-1\,0\,0\,0\,0), (0\,1\,0\,-1\,0), (-1\,1\,1\,-1\,0).$

$D_K = -30976, \omega = i, f(t) = t^3 - t^2 + (2 - \omega)t - 1$
Solutions: $(x_1, x_2, y_0, y_1, y_2) = (0\,0\,-2\,1\,-1), (0\,0\,-1\,1\,-1), (-1\,0\,0\,0\,0),$
$(0\,0\,0\,-1\,0), (0\,1\,-2\,1\,-1), (-1\,1\,-1\,0\,0), (-1\,1\,0\,0\,0), (-1\,1\,0\,0\,1).$

$D_K = -31347, \omega = (1 + i\sqrt{3})/2, f(t) = t^3 + (-1 - \omega)t^2 + 3\omega t - \omega$
Solutions: $(x_1, x_2, y_0, y_1, y_2) = (-2\,0\,0\,1\,-1), (-1\,0\,-2\,1\,-1), (-1\,0\,-1\,1\,-1),$
$(1\,0\,0\,1\,-1), (-1\,0\,0\,0\,0), (-1\,0\,-1\,1\,0), (0\,0\,1\,-1\,0), (-2\,1\,0\,1\,-1),$
$(-1\,1\,0\,0\,-1), (-1\,1\,2\,-1\,0), (0\,1\,2\,-1\,0).$

$D_K = -33856, \omega = i, f(t) = t^3 + t - \omega$
Solutions: $(x_1, x_2, y_0, y_1, y_2) = (0\,0\,-1\,0\,-1), (0\,0\,0\,0\,-1), (1\,0\,-1\,0\,-1),$
$(-1\,0\,0\,0\,0), (-1\,1\,0\,0\,1), (1\,1\,0\,0\,-1).$

$D_K = -34371, \omega = (1 + i\sqrt{3})/2,$
$f(t) = t^3 + (-1 - \omega)t^2 + (-1 + 4\omega)t + (2 - \omega)$
Solutions: $(x_1, x_2, y_0, y_1, y_2) = (0\,0\,-2\,2\,-1), (0\,0\,1\,-1\,0), (-2\,1\,-2\,3\,-2),$
$(-2\,1\,0\,2\,-1), (-1\,1\,0\,1\,-1), (-1\,1\,2\,0\,0), (-1\,1\,3\,0\,0), (-3\,2\,0\,4\,-2).$

$D_K = -34992, \omega = (1 + i\sqrt{3})/2, f(t) = t^3 + (-1 - \omega)t^2 + 3\omega t + (1 - 2\omega)$
Solutions: $(x_1, x_2, y_0, y_1, y_2) = (0\,0\,1\,-1\,0), (-1\,1\,0\,0\,-1), (-1\,1\,0\,1\,-1).$

$D_K = -36963, \omega = (1 + i\sqrt{3})/2, f(t) = t^3 + (-1 - \omega)t^2 + \omega t + (-1 + \omega)$
Solutions: $(x_1, x_2, y_0, y_1, y_2) = (-1\,0\,0\,1\,-1), (-1\,0\,-1\,2\,-1), (-1\,0\,0\,0\,0),$
$(-1\,0\,0\,1\,0), (0\,0\,1\,-1\,0), (-2\,1\,0\,1\,-1), (-1\,1\,0\,1\,-1), (-1\,1\,0\,0\,0),$
$(0\,1\,0\,-1\,0).$

$D_K = -40203, \omega = (1 + i\sqrt{3})/2,$
$f(t) = t^3 + (-1 - \omega)t^2 + (-2 + 3\omega)t + (2 - \omega)$
Solutions: $(x_1, x_2, y_0, y_1, y_2) = (-1\,0\,0\,1\,-1), (0\,0\,0\,1\,-1), (-1\,0\,0\,1\,0),$
$(0\,0\,1\,-1\,0), (-1\,1\,2\,0\,0).$

$D_K = -41472, \omega = i\sqrt{2}, f(t) = t^3 + (1 - \omega)t - 1$
Solutions: $(x_1, x_2, y_0, y_1, y_2) = (1\,0\,-1\,0\,-1), (-1\,0\,0\,0\,0), (1\,1\,-1\,0\,-1),$
$(0\,1\,-1\,0\,0), (-1\,1\,0\,1\,1), (0\,1\,0\,1\,1).$

$D_K = -41823, \omega = (1 + i\sqrt{3})/2, f(t) = t^3 - t^2 + (5 - 5\omega)t + (-6 + 2\omega)$
Solutions: $(x_1, x_2, y_0, y_1, y_2) = (1\,0\,1\,-1\,-1), (3\,1\,-1\,0\,-3), (0\,1\,-3\,1\,-1),$
$(1\,1\,-3\,0\,-1), (-2\,1\,-5\,2\,1), (0\,2\,-7\,2\,-1).$

References

[Ar74] G. Archinard, *Extensions cubiques cycliques de $\mathbb{Q}$ dont l'annaeau des entiers est monogéne*, Enseignement Math., **20**(1974), 179–203.

[Ba90] A. Baker, *Transcendental Number Theory*, Cambridge, 1990.

[BD69] A. Baker and H. Davenport, *The equations $3x^2 - 2 = y^2$ and $8x^2 - 7 = z^2$*, Quart. J. Math. Oxford, **20**(1969), 129–137.

[BW93] A. Baker and G. Wüstholz, *Logarithmic forms and group varieties*, J. Reine Angew. Math., **442**(1993), 19–62.

[BMO90] A.M. Bergé, J. Martinet, M. Olivier, *The computation of sextic fields with a quadratic subfield*, Math. Comput., **54**(1990), 869–884.

[BH96] Y. Bilu and G. Hanrot, *Solving Thue equations of high degree*, J.Number Theory, **60**(1996), 373–392.

[BM71] B.J. Birch and J.R. Merriman, *Finiteness theorems for binary forms with given discriminant*, Proc. London Math. Soc., **24**(1972), 385–394.

[Br85] A. Bremner, *Integral generators in a certain quartic field and related diophantine equations*, Michigan Math. J., **32**(1985), 295–319.

[Br88] A. Bremner, *On power bases in cyclotomic fields*, J. Number Theory, **28**(1988), 288–298.

[BF89] J. Buchmann and D. Ford, *On the computation of totally real quartic fields of small discriminant*, Math. Comput., **52**(1989), 161–174.

[BFP93]   J. Buchmann, D. Ford and M. Pohst, *Enumeration of quartic fields of small discriminant*, Math. Comput., **61**(1993), 873–879.

[BGy96a]   Y. Bugeaud and K. Győry, *Bounds for the solutions of unit equations*, Acta Arith., **74**(1996), 67–80.

[BGy96b]   Y. Bugeaud and K. Győry, *Bounds for the solutions of Thue–Mahler equations and norm form equations*, Acta Arith., **74**(1996), 273–292.

[CG88]   B.W. Char, K.O. Geddes, G.H. Gonnet, M.B. Monagan, S.M. Watt (eds.), *MAPLE, Reference Manual*, Watcom Publications, Waterloo, Canada, 1988.

[Co93]   H. Cohen, *A Course in Computational Algebraic Number Theory*, Springer-Verlag, 1993.

[DF97]   M. Daberkow, C. Fieker, J. Klüners, M. Pohst, K. Roegner and K. Wildanger, *KANT V4* , J. Symbolic Comput., **24**(1997), 267–283.

[Da91]   H. Darmon, *Note on a polynomial of Emma Lehmer*, Math. Comput., **56**(1991), 795–800.

[Ded878]   R. Dedekind, *Über Zusammenhang zwischen der Theorie der Ideale und der Theorie der höhere Kongruenzen*, Abh. König. Ges. der Wissen. zu Göttingen, **23**(1878), 1–23.

[Del30]   B.N. Delone, *Über die Darstellung der Zahlen durch die binären kubischen Formen von negativer Diskriminante*, Math. Z., **31**(1930), 1–26.

[DK77]   D.S. Dummy and H. Kisilevsky, *Indices in cyclic cubic fields*, in "Number Theory and Algebra", Academic Press, 1977, pp. 29–42.

[En30]   H.T. Engstrom, *On the common index divisors of an algebraic field*, Trans. Amer. Math. Soc., **32**(1930), 223–237.

[EMT85]   V. Ennola, S. Mäki and R. Turunen, *On real cyclic sextic fields*, Math. Comput., **45**(1985), 591–611.

[FP85]   U. Fincke and M. Pohst, *Improved methods for calculating vectors of short length in a lattice, including a complexity analysis*, Math. Comput., **44**(1985), 463–471.

[Fo91]   D. Ford, *Enumeration of totally complex quartic fields of small discriminant*, in "Computational Number Theory", ed. by A. Pethő, M.E. Pohst, H.C. Williams and H.G. Zimmer, Walter de Gruyter, Berlin–New York 1991, pp. 129–138.

[Ga88]   I. Gaál, *On the resolution of inhomogeneous norm form equations in two dominating variables*, Math. Comput., **51**(1988), 359–373.

[Ga91]   I. Gaál, *On the resolution of some diophantine equations,* in "Computational Number Theory", ed. by A. Pethő, M.E. Pohst, H.C. Williams and H.G. Zimmer, Walter de Gruyter, Berlin–New York, 1991, pp. 261–280.

[Ga93]   I. Gaál, *Power integral bases in orders of families of quartic fields,* Publ. Math. (Debrecen), **42** (1993), 253–263.

[Ga95]   I. Gaál, *Computing elements of given index in totally complex cyclic sextic fields,* J. Symbolic Comput., **20**(1995), 61–69.

[Ga96a]   I. Gaál, *Computing all power integral bases in orders of totally real cyclic sextic number fields,* Math. Comput., **65**(1996), 801–822.

[Ga96b]   I. Gaál, *Application of Thue equations to computing power integral bases in algebraic number fields,* Proc. Conf. ANTS II, Talence, France, 1996. Lecture Notes in Computer Science 1122, Springer 1996, pp. 151–155.

[Ga98a]   I. Gaál, *Power integral bases in composites of number fields,* Canad. Math. Bulletin, **41**(1998), 158–165.

[Ga98b]   I. Gaál, *Computing power integral bases in algebraic number fields,* in "Number Theory", ed. by K. Győry, A. Pethő and V.T. Sós, Walter de Gruyter, Berlin-New York, 1998, pp. 243–254.

[Ga99]   I. Gaál, *Power integral bases in algebraic number fields,* Ann. Univ. Sci. Budapestiensis R. Eötvös Nom., Sect. Computatorica, **18**(1999), 61–87.

[Ga00a]   I. Gaál, *Computing power integral bases in algebraic number fields II,* in "Algebraic number theory and diophantine analysis", ed. by F. Halter-Koch and R.F. Tichy, Walter de Gruyter, Berlin-New York, 2000, pp. 153–161.

[Ga00b]   I. Gaál, *Solving index form equations in fields of degree nine with cubic subfields,* J. Symbolic Comput., **30**(2000), 181–193.

[Ga00c]   I. Gaál, *An efficient algorithm for the explicit resolution of norm form equations,* Publ. Math. (Debrecen), **56**(2000), 375–390.

[Ga01]   I. Gaál, *Power integral bases in cubic relative extensions,* Experimental Math., **10**(2001), 133–139.

[GGy99]   I. Gaál and K. Győry, *Index form equations in quintic fields,* Acta Arith., **89**(1999), 379–396.

[GNy01]   I. Gaál and G. Nyul, *Computing all monogeneous dihedral quartic extensions of a quadratic field,* J. Theorie Nombres Bordeaux, **13**(2001), 137–142.

[GOP01] I. Gaál, P. Olajos and M. Pohst, *Power integer bases in orders of composite fields*, Experimental Math., to appear.

[GPP91a] I. Gaál, A. Pethő and M. Pohst, *On the resolution of index form equations in biquadratic number fields, I*, J. Number Theory, **38**(1991), 18–34.

[GPP91b] I. Gaál, A. Pethő and M. Pohst, *On the resolution of index form equations in biquadratic number fields, II*, J. Number Theory, **38**(1991), 35–51.

[GPP91c] I. Gaál, A. Pethő and M. Pohst, *On the indices of biquadratic number fields having Galois group V₄*, Arch. Math., **57**(1991), 357–361.

[GPP91d] I. Gaál, A. Pethő and M. Pohst, *On the resolution of index form equations*, Proc. of the 1991 International Symposium on Symbolic and Algebraic Computation, ed. by Stephen M. Watt, ACM Press, 1991, pp. 185–186.

[GPP93] I. Gaál, A. Pethő and M. Pohst, *On the resolution of index form equations in quartic number fields*, J. Symbolic Comput., **16**(1993), 563–584.

[GPP94] I. Gaál, A. Pethő and M. Pohst, *On the resolution of index form equations in dihedral number fields*, Experimental Math., **3**(1994), 245–254.

[GPP95] I. Gaál, A. Pethő and M. Pohst, *On the resolution of index form equations in biquadratic number fields, III. The bicyclic biquadratic case*, J. Number Theory, **53**(1995), 100–114.

[GPP96] I. Gaál, A. Pethő and M. Pohst, *Simultaneous representation of integers by a pair of ternary quadratic forms — with an application to index form equations in quartic number fields*, J. Number Theory, **57**(1996), 90–104.

[GP96] I. Gaál and M. Pohst, *On the resolution of index form equations in sextic fields with an imaginary quadratic subfield*, J. Symbolic Comput., **22**(1996), 425–434.

[GP97] I. Gaál and M. Pohst, *Power integral bases in a parametric family of totally real quintics*, Math. Comput., **66**(1997), 1689–1696.

[GP00] I. Gaál and M. Pohst, *On the resolution of index form equations in relative quartic extensions*, J. Number Theory, **85**(2000), 201–219.

[GP01] I. Gaál and M. Pohst, *On the resolution of relative Thue equations*, Math. Comput., **71**(2002), 429–440.

[GS89] I. Gaál and N. Schulte, *Computing all power integral bases of cubic number fields*, Math. Comput., **53**(1989), 689–696.

[Gr73]   M.N. Gras, *Sur les corps cubiques cycliques dont l'anneau des entiers monogéne*, Publ. Math. Fac. Sci. Besancon, 1973.

[Gr75]   M.N. Gras, *Lien entre le groupe des unites et la monogéneite des corps cubiques cycliques*, Theorie des Nombres Besancon, Années 1975–76.

[Gr79]   M.N. Gras, *Z–bases d'entiers* 1, ϑ, ϑ^2, ϑ^3 *dans les extensions cycliques de degre 4 de* $\mathbb{Q}$, Theorie des Nombres Besancon, Années 1979–1980 et 1980–1981.

[Gr83]   M.N. Gras, *Non monogénéité de l'anneau des entiers de certaines extensions abeliennes de* $\mathbb{Q}$, Publ. Math. Fac. Sci. Besancon, Theor. Nombres, 1983–1984, 25 pp.

[Gr86]   M.N. Gras, *Non monogénéité de l'anneau des entiers des extensions cycliques de* $\mathbb{Q}$ *de degré premier* $l \geq 5$, J. Number Theory, **23**(1986), 347–353.

[GT95]   M.N. Gras and F. Tanoe, *Corps biquadratiques monogénes*, Manuscripta Math., **86**(1995), 63–79.

[Gy76]   K. Győry, *Sur les polynomes a coefficients entiers et de discriminant donne, III*, Publ. Math. (Debrecen), **23**(1976), 141–165.

[Gy81]   K. Győry, *On the representation of integers by decomposable forms in several variables*, Publ. Math. (Debrecen), **28**(1981), 89–98.

[Gy98]   K. Győry, *Bounds for the solutions of decomposable form equations*, Publ. Math. (Debrecen), **52**(1998), 1–31.

[Gy00]   K. Győry, *Discriminant form and index form equations*, in "Algebraic number theory and diophantine analysis", ed. by F. Halter-Koch and R.F. Tichy, Walter de Gruyter, Berlin-New York, 2000, pp. 191–214.

[Ha63]   H. Hasse, *Zahlentheorie*, Akademie-Verlag, Berlin, 1963.

[He08]   K. Hensel, *Theorie der algebraischen Zahlen*, Teubner Verlag, Leipzig-Berlin, 1908.

[Ja01]   I. Járási, *Power integral bases in sextic fields with a cubic subfield*, Acta Sci. Math. Szeged, to appear.

[Ka99]   A.C. Kable, *Power integral bases in dihedral quartic fields*, J. Number Theory, **76**(1999), 120–129.

[KW89]   L.C. Kappe and B. Warren, *An elementary test for the Galois group of a quartic polynomial*, Amer. Math. Monthly, **96**(1989), 133–137.

[Kl95]   M. Klebel, *Zur Theorie der Potenzganzheitsbases bei relativ galoisschen Zahlkörpern*, Thesis, Univ. Augsburg, 1995.

176   References

[Koc66]   R. Kochendörfer, *Einführung in die Algebra*, VEB Deutscher Verlag der Wissenschaften, Berlin, 1966.

[Kop94]   D. Koppenhöfer, *Über projektive Darstellungen von Algebren kleinen Ranges*, Thesis, Univ. Tübingen, 1994.

[Kop95]   D. Koppenhöfer, *Determining the monogenity of a quartic number field*, Math. Nachrichten, **172**(1995), 191–198.

[KP91]   B. Kovács and A. Pethő, *Number systems in integral domains, especially in orders of algebraic number fields*, Acta Sci. Math., **55**(1991), 287–299.

[Kr882]   L. Kronecker, *Grundzüge einer arithmetischen Theorie der algebraischen Größen*, J. Reine Angew. Math., **92**(1882), 1–122.

[Le88]   E. Lehmer, *Connection between Gaussian periods and cyclic units*, Math. Comput., **50**(1988), 535–541.

[LLL82]   A.K. Lenstra, H.W. Lenstra Jr. and L. Lovász, *Factoring polynomials with rational coefficients*, Math. Ann., **261**(1982), 515–534.

[Le99]   F. Leprevost, *Sur certaines surfaces elliptiques et courbes elliptiques de Mordell de rang $\geq$ 1 associées á des discriminants de polynômes cubiques ou quartiques*, J. Number Theory, **78**(1999), 149–165.

[LP95]   G. Lettl and A. Pethő, *Complete solution of a family of quartic Thue equations*, Abh. Math. Sem. Univ. Hamburg, **65**(1995), 365–383.

[LPV98]   G. Lettl, A. Pethő, P. Voutier, *On the arithmetic of simplest sextic fields and related Thue equations*, in "Number Theory", ed. by K. Győry, A. Pethő, V.T. Sós, Walter de Gruyter, Berlin-New York, 1998, pp. 331–348.

[Ma80]   S. Mäki, *The Determination of Units in Real Cyclic Sextic Fields*, Lecture Notes in Mathematics, No. 797, Springer Verlag, Berlin–Heidelberg–New York, 1980.

[MS93]   J.R. Merriman and N.P. Smart, *The calculation of all algebraic integers of degree 3 with discriminant a product of 2 and 3 only*, Publ. Math. (Debrecen), **43**(1993), 195–205.

[Mi93]   M. Mignotte, *Verification of a conjecture of E. Thomas*, J. Number Theory, **44**(1993), 172–177.

[MT91]   M. Mignotte and N. Tzanakis, *On a family of cubics*, J. Number Theory, **39**(1991), 41–49.

[Mo69]   L.J. Mordell, *Diophantine Equations*, Academic Press, New York–London, 1969.

[Nag30]   T. Nagell, *Zur Theorie der kubischen Irrationalitäten*, Acta Math., **55**(1930), 33–65.

[Nag67]   T. Nagell, *Sur les discriminants des nombres algébriques*, Arkiv för Mat., **7**(1967), 265–282.

[Nag68]   T. Nagell, *Quelques propriétés des nombres algébriques du quatrième degré*, Arkiv för Mat., **7**(1968), 517–525.

[Nak83]   T. Nakahara, *On the indices and integral bases of non–cyclic but abelian biquadratic fields*, Arch. Math., **41**(1983), 504–508.

[Nak87]   T. Nakahara, *On the minimum index of a cyclic quartic field*, Arch. Math., **48**(1987), 322–325.

[Nak93]   T. Nakahara, *A simple proof for non–monogenesis of the rings of integers in some cyclic fields*, in "Advances in Number Theory", ed. by F.Q. Gouvéa and N. Yui, Clarendon Press, Oxford, 1993, pp. 167–173.

[Nark74]   W. Narkiewicz, *Elementary and Analytic Theory of Algebraic Numbers*, Second Edition, Springer, 1974.

[Nart85]   E. Nart, *On the index of a number field*, Trans. Amer. Math. Soc., **289**(1985), 171–183.

[Ne91]   I. Nemes, *On the solution of the diophantine equation $G_n = P(x)$ with sieve algorithm*, in "Computational Number Theory" ed. by A. Pethő, M.E. Pohst, H.C. Williams, H.G. Zimmer, Walter de Gruyter, Berlin–New York, 1991, pp. 303–311.

[Ny01]   G. Nyul, *Power integral bases in totally complex biquadratic number fields*, Acta Acad. Paed. Aegriensis, Sect. Math., **28**(2001), 79–86.

[Ola01]   P. Olajos, *Power integral bases in a parametric family of sextic fields*, Publ. Math. (Debrecen), **58**(2001), 779–790.

[Oli89]   M. Olivier, (1989). *Corps sextiques contenant un corps quadratique (I).*, Séminaire de Théorie des Nombres Bordeaux, **1**(1989), 205–250.

[Or28]   O. Ore, *Newtonsche Polygone in der Theorie der algebraischen Körper*, Math. Ann., **99**(1928), 84–117.

[Pe87]   A. Pethő, *On the resolution of Thue inequalities*, J. Symbolic Comput., **4**(1987), 103–109.

[Pe99]   A. Pethő, *Algebraische Algorithmen*, Vieweg Verlag, Braunschweig-Wiesbaden, 1999.

[Pe01]   A. Pethő, *Index form surfaces and construction of elliptic curves over large fields*, manuscript.

178   References

[PS87]   A. Pethő and R. Schulenberg, *Effektives Lösen von Thue Gleichungen*, Publ. Math. (Debrecen), **34**(1987), 189–196.

[Pi88]   R.G.E. Pinch, *Simultaneous Pellian equations*, Math. Proc. Cambridge Philos. Soc., **103**(1988), 35–46.

[Po93]   M. Pohst, *Computational Algebraic Number Theory*, DMV Seminar Band 21, Birkhäuser, 1993.

[Po00]   M. Pohst, *On Legendre's equation over number fields*, Publ. Math. (Debrecen), **56**(2000), 535–546.

[PZ89]   M. Pohst and H. Zassenhaus, *Algorithmic algebraic number theory*, Cambridge University Press, 1989.

[Ro98]   L. Robertson, *Power bases for cyclotomic integer rings*, J. Number Theory, **69**(1998), 98–118.

[Ro01]   L. Robertson, *Power bases for 2-power cyclotomic fields*, J. Number Theory, **88**(2001), 196–209.

[Schl77] H.P. Schlickewei, *On norm form equations*, J. Number Theory, **9**(1977), 370–380.

[Schm72] W.M. Schmidt, *Norm form equations*, Ann. Math., **96**(1972), 526–551.

[SW88]   R. Schoof and L. Washington, *Quintic polynomials and real cyclotomic fields with large class numbers*, Math. Comput., **50**(1988), 543–556.

[Schu89] N. Schulte, *Indexgleichungen in kubischen Zahlkörpern*, Diplomarbeit, Heinrich–Heine Universität, Düsseldorf, 1989.

[Schu91] N. Schulte, *Index form equations in cubic number fields*, in "Computational Number Theory", ed. by A. Pethő, M.E. Pohst, H.C. Williams and H.G. Zimmer, Walter de Gruyter, Berlin–New York, 1991, pp. 281–287.

[Sh74]   D. Shanks, *The simplest cubic fields*, Math. Comput., **28**(1974), 1137–1152.

[Si79]   C.L. Siegel, *Normen algebraischer Zahlen*, in "C.L.Siegel, *Gesammelte Abhandlungen*", Band IV, Springer Verlag, 1979, pp. 250–268.

[Sl82]   J. Sliwa, *On the nonessential discriminant divisor of an algebraic number field*, Acta Arith., **42**(1982), 57–72.

[Sm93]   N.P. Smart, *Solving a quartic discriminant form equation*, Publ. Math. (Debrecen), **43**(1993), 29–39.

[Sm95]   N.P. Smart, *The solution of triangularly connected decomposable form equations*, Math. Comput., **64**(1995), 818–840.

[Sm96]   N.P. Smart, *Solving discriminant form equations via unit equations*, J. Symbolic Comput., **21**(1996), 367–374.

[Sm97]   N.P. Smart, *Thue and Thue-Mahler equations over rings of integers*, J. London Math. Soc. (2), **56**(1997), 455–462.

[Sm98]   N.P. Smart, *The Algorithmic Resolution of Diophantine Equations*, London Math. Soc., Student Texts 41, Cambridge University Press, 1998.

[Sp74]   V.G. Sprindžuk, *Representation of numbers by the norm forms with two dominating variables*, J. Number Theory, **6**(1974), 481–486.

[Sy00]   Syed Inayat Ali Shah, *Monogenesis of the ring of integers in a cyclic sextic field of a prime conductor*, Rep. Fac. Sci. Engrg. Saga Univ. Math., **29**(2000), 1–10.

[The93]   J.D. Thérond, *Extensions cycliques cubiques monogénes de l'anneau des entiers d'un corps quadratique imaginaire*, Arch. Math., **61**(1993), 348–361.

[The95]   J.D. Thérond, *Extensions cycliques cubiques monogénes de l'anneau des entiers d'un corps quadratique*, Arch. Math., **64**(1995), 216–229.

[Tho90]   E. Thomas, *Complete solutions to a family of cubic diophantine equations*, J. Number Theory, **34**(1990), 235–250.

[Thu09]   A. Thue, *Über Annäherungswerte algebraischer Zahlen*, J. Reine Angew. Math., **135**(1909), 284–305.

[TW89]   N. Tzanakis and B.M.M. de Weger, *On the practical solution of the Thue equation*, J. Number Theory, **31**(1989), 99–132.

[TW92]   N. Tzanakis and B.M.M. de Weger, *How to explicitely solve a Thue Mahler equation*, Compositio Math., **84**(1992), 223–288.

[We89]   B.M.M. de Weger, *Algorithms for Diophantine Equations*, CWI Tract 65, Amsterdam, 1989.

[We95]   B.M.M. de Weger, *A Thue equation with quadratic integers as variables*, Math. Comput., **64**(1995), 855–861.

[Wi97]   K. Wildanger, *Über das Lösen von Einheiten- und Indexformgleichungen in algebraischen Zahlkörpern mit einer Anwendung auf die Bestimmung aller ganzen Punkte einer Mordellschen Kurve*, Thesis, Technical University, Berlin, 1997.

[Wi00]   K. Wildanger, *Über das Lösen von Einheiten- und Indexformgleichungen in algebraischen Zahlkörpern*, J. Number Theory, **82**(2000), 188–224.

[Will70]   K.S. Williams, *Integers of biquadratic fields*, Canad. Math. Bull., **13**(1970), 519–526.

[Zi81]   R. Zimmert, *Ideale kleiner Norm in Idealklassen und eine Regulatorabschätzung*, Invent. Math., **62**(1981), 367–380.

Author Index

Subject Index